Nanotechnology in Catalysis
Volume 2

Nanostructure Science and Technology

Series Editor: David J. Lockwood, FRSC
National Research Council of Canada
Ottawa, Ontario, Canada

Current volumes in this series:

Alternative Lithography: Unleashing the Potentials of Nanotechnology
Edited by Clivia M. Sotomayor Torres

Nanoparticles: Building Blocks for Nanotechnology
Edited by Vincent Rotello

Nanostructured Catalysts
Edited by Susannah L. Scott, Cathleen M. Crudden, and Christopher W. Jones

Nanotechnology in Catalysis, Volumes 1 and 2
Edited by Bing Zhou, Sophie Hermans, and Gabor A. Somorjai

Polyoxometalate Chemistry for Nano-Composite Design
Edited by Toshihiro Yamase and Michael T. Pope

Self-Assembled Nanostructures
Jin Z. Zhang, Zhong-lin Wang, Jun Liu, Shaowei Chen, and Gang-yu Liu

Semiconductor Nanocrystals: From Basic Principles to Applications
Edited by Alexander L. Efros, David J. Lockwood, and Leonid Tsybeskov

A Continuation Order Plan is available for this series. A continuation order will bring delivery of each new volume immediately upon publication. Volumes are billed only upon actual shipment. For further information please contact the publisher.

Nanotechnology in Catalysis
Volume 2

Edited by

Bing Zhou
Headwaters NanoKinetix, Inc.
Lawrenceville, New Jersey

Sophie Hermans
Catholic University of Louvain
Louvain-La-Neuve, Belgium

and

Gabor A. Somorjai
University of California
Berkeley, California

Kluwer Academic / Plenum Publishers
New York   Boston   Dordrecht   London   Moscow

012754833

CHEM

Library of Congress Cataloging-in-Publication Data

Nanotechnology in catalysis/edited by Bing Zhou, Sophie Hermans, Gabor A. Somorjai.
    p.   cm. — (Nanoscience science and technology)
    Includes bibliographical references and index.
    ISBN 0-306-48323-8
    1. Catalysts—Congresses. 2. Nanotechnology—Congresses. 3. Nanostructure
materials—Congresses. I. Zhou, Bing, Dr. II. Hermans, Sophie, Dr. III. Somorjai, Gabor A., Prof.
IV. American Chemical Society. Meeting (221st: 2001: San Diego, Calif.) V. American
Chemical Society. Meeting (224th: 2002: Boston, Mass.) VI. Series.

TP159.C3N37 2004
660′.2995—dc22

2003064027

ISBN 0-306-48323-8

©2004 Kluwer Academic/Plenum Publishers, New York
233 Spring Street, New York, New York 10013

http://www.kluweronline.com

10 9 8 7 6 5 4 3 2 1

A C.I.P. record for this book is available from the Library of Congress

Permissions for books published in Europe: *permissions@wkap.nl*
Permissions for books published in the United States of America: *permissions@wkap.com*

Printed in the United States of America

Contributors

Volume II

R.T. Baker, University of Dundee, Dundee, United Kingdom

J.M. Basset, Laboratoire de Chimie Organometallique de Surface, Villeurbanne Cedex, France

D.C. Bazin, Universite Paris XI, Orsay Cedex, France

S. Bernal, Universidad de Cadiz, Cadiz, Spain

J.J. Calvino, Universidad de Cadiz, Cadiz, Spain

J.P. Candy, Laboratoire de Chimie Organometallique de Surface, Villeurbanne Cedex, France

C. Coperet, Laboratoire de Chimie Organometallique de Surface, Villeurbanne Cedex, France

L. Guczi, Institute of Isotope and Surface Chemistry, Hungarian Academy of Sciences, Budapest, Hungary

M.R. Hajaligol, Philip Morris USA, Richmond, Virginia

J.C. Hatfield, Dow Chemical Company, South Charleston, West Virginia

F. Lefebvre, Laboratoire de Chimie Organometallique de Surface, Villeurbanne Cedex, France

C. Li, Dalian Institute of Chemical Physics, Dalian, China

P. Li, Philip Morris USA, Richmond, Virginia

Z.L. Li, Dalian Institute of Chemical Physics, Dalian, China

C.H. Liang, Dalian Institute of Chemical Physics, Dalian, China

J. Liu, Monsanto Company., St. Louis, Missouri

C.Lopez-Cartes, Instituto de Ciencia de Materiales de Sevilla, Sevilla, Spain

D.E. Miser, Philip Morris USA, Richmond, Virginia

Z. Paszti, Institute of Isotope and Surface Chemistry, Hungarian Academy of Sciences, Budapest, Hungary

J.A. Perez-Omil, Universidad de Cadiz, Cadiz, Spain

G. Petö, Institute of Technical Physics and Materials Science, Hungarian Academy of Sciences, Budapest, Hungary

E.A Quandrelli, Laboratoire de Chimie Organometallique de Surface, Villeurbanne Cedex, France

J.S. Qui, Dalian University of Technology, Dalian, China

F. Rasouli, Philip Morris USA, Richmond, Virginia

E.J. Shin, Philip Morris USA, Richmond, Virginia

M.L. Tulchinsky, Dow Chemical Company, South Charleston, West Virginia

Z.B. Wei, Dalian Institute of Chemical Physics, Dalian, China

Q. Xin, Dalian Institute of Chemical Physics, Dalian, China

Preface

Catalysts, heterogeneous, homogeneous and enzyme, are usually nanoparticles. These are of vital for the functioning of the human body, for photosynthesis, and for producing fuels and chemicals in the petroleum and chemical industries. Interest in nanoscience and in nanotechnology in recent years focused attention on the opportunity to develop catalysts that exhibit 100% selectivity for a desired product, thus removing byproducts and eliminating waste. This type of selective process is often called green chemistry or green technology.

This book is mainly based on the first and second symposia on Nanotechnology in Catalysis which were held in spring 2001 at the ACS 221st National Meeting in San Diego, CA, and in fall 2002 at the ACS 224th National Meeting in Boston, MA, respectively. We also extended our invitation to those who did not attend the meetings to contribute chapters where we saw a need to round out the scope of the topic. All chapters were peer-reviewed prior to final acceptance. We believe that the additional chapters and the peer-review significantly improved the quality of the book.

In the summer of 2000 when we first proposed to organize a symposium on Nanotechnology in Catalysis to the ACS Secretariat of Catalysis and Surface Science (CATL), we received strong support from Dr. Nancy B. Jackson, then General Secretary of CATL. The symposium was enthusiastically received by the catalysis community. On the first day of the symposium, the conference room could not hold all the attendees. People were standing behind the last row of chairs or at the door to listen to the speakers.

Nanotechnology has become an important area globally. US Government spending on nanotechnology over the last two years is estimated at $2 billion. In The United States, legislation passed by the House of Representatives in April 2003 authorized $2.135 billion in federal research money for nanotechnology research and development over the next 3 years (Nano/Bio Convergence News, Vol.1, No.9, May 2003). The National Science Foundation (NSF) forecasts that the market value of nano products and services will reach $1 trillion by 2015 (NSF: Societal Implications of nanoscience and nanotechnology, March 2001). Similarly, the European Commission proposal for the 6th Framework Programme (2002 - 2006) contains a strong focus on nanotechnology. Out of a total proposed funding of 17.5 € billion, 1.3 € billion would be devoted to *"a priority thematic area of research on nanotechnology, knowledge-based materials and new industrial processes"* (source: www.cordis.lu, website of the European Commission).

Catalysis research and catalyst-based technologies have been at the heart of nanotechnology for many years. Nanotechnology is about manipulating and making materials at the atomic and molecular level. The development of supported noble metal catalysts in the 1950s aimed at reducing costs for large commercial applications resulted in catalysts with noble metal particle of sizes less than 10 nm, which by today's standard are nanomaterials. Zeolite catalysts, discovered in the late 1960s, are another example. By deliberate design and preparation of the catalyst structure at the atomic and molecular level, researchers at Mobil Oil Co. were able to synthesize zeolites such as ZSM-5, a nanostructured crystalline material with a 10-atom ring and pore size of 0.45-0.6 nm, enabling the control of selectivity for petrochemical processes at a molecular level. Such nanomaterial catalysts revolutionized the petrochemical industry. Today, zeolite catalysts are used in processing over 7 billion barrels of petroleum and also many chemicals annually.

Research and development in the catalysis field have been in the nanometer scale since then. Recent developments of modern tools to characterize materials in nano or subnano scale provide insight for understanding and improving the existing catalysts, and clues for designing new nanomaterials for better catalysts.

The papers of this book reflect some of the frontier areas of nanoscience and nanotechnology to fabricate and characterize catalysts and carry out reaction studies to prove their selectivity and activity. This field of application of nanotechnology for the development of green catalysts is likely to grow rapidly during the next decade. This book hopes to contribute to the evolution of nanotechnology in this direction. The book is also a summary of updated advances and breakthroughs achieved worldwide by researchers in nanotechnology in the catalysis area. It is a difficult task to cover all aspects of such a dynamic research area. Also, there is no clear cut way to assign each contribution to well-defined topics. However, to facilitate comprehension of the advances in the field, the papers are organized into the following five sections.

Section I provides an overview of the fundamental understanding of catalysis and nanoscience. The evolution of the field of catalysis and its relation to nanoscience and nanotechnology are discussed. The authors describe the fabrication of 2- and 3-dimensional nanoparticle catalysts with controlled structures using electron beam and photo-lithography, providing the insight for possible catalyst fabrication in the 21st Century.

Section II focuses on nanoparticle and nanocluster catalysts. In this section, Chapters 2 to 6 describe the recent developments in synthesis and characterization of nanoparticle or nanocluster catalysts. Chapters 7 and 8 discuss the use of nanoparticle catalysts to grow carbon nanomaterials. The last part of this section (Chapters 9, 10 and 11) is devoted to noble metal nanoparticle materials as electrocatalysts, which are of vital importance for energy generation by fuel cells in the future.

Section III summarizes the recent advances in nanoporous materials as catalysts or catalytic supports. By controlling the structure of such materials at the atomic and molecular level, shape-selective and regio-specific catalysts are developed. The exceptional high selectivity of such nanoporous materials toward specific desired products has a potential to reduce or eliminate waste production. The chapters in this section provide clues for the new generation catalysts of the 21st Century, which may lead to green chemistry or green technology.

Section IV concentrates on the characterization and understanding of nanostructured catalysts and their properties by using modern tools and recently developed theories. Chapters 18 and 19 present how to use advanced and high-resolution electron microscopy to obtain detailed structural information at the nano-scale for catalyst development. New concepts and theories of characterization and understanding of heterogeneous catalysis at the nano-scale are discussed in Chapter 20 to 22.

Section V presents three examples of new nanomaterials as catalysts or supports. Researchers from the Dow Chemical Company explore the use of nanoscale dendrimers as hydroformylation catalysts. A study from Philip Morris examines the catalytic effect of nanoparticle iron oxide on carbon monoxide and biomass compounds. And finally, a paper from the Dalian Institute of Chemical Physics in China discusses the use of graphitic nanofilaments as a superior catalyst support for ammonia synthesis.

In the spring of 2002 when we discussed the organization of the second symposium on Nanotechnology in Catalysis at the Catalysis Society of Metropolitan New York, both Professor Israel Wachs from Lehigh University and Dr. Gary McVicker from ExxonMobil noted that catalysis is an area that has moved one-step ahead of other areas in nanotechnology development. We are currently studying the catalyst materials at a subnanometer level. This provides an insight for how far we have advanced in the metered scale in catalysis. We have passed the nanometer level!

The first and second symposia on Nanotechnology in Catalysis were successful. We believe that their success and popularity are reflected in this book, which provides information on what has been done, and hopefully, an insight into what may happen in the future.

Gabor A. Somorjai
Bing Zhou
Sophie Hermans

Acknowlededgments

We are grateful to many people who have assisted and supported us in so many ways during the editing of this book. Without their involvement, the book would not have reached readers.

The suggestion from Dr. Kenneth Howell, Senior Editor at Kluwer Academic/Plenum Publishers, to publish the symposia on Nanotechnology in Catalysis as a book was a pleasant surprise to us. His initiation, dedication and persistence over last year helped us to overcome many obstacles and made the book project possible. His guidance on many detailed and time-consuming issues of the book was precise and led us through the difficult times. The vision of Kluwer Academic/Plenum Publishers in promoting science and technology is also acknowledged wholeheartedly.

The authors of each chapter in this book are acknowledged for their due diligence. Their contribution was essential to obtaining a high quality book. Many of the chapters have updated summaries of recent breakthroughs and developments made by the authors. We wish to thank all of them for their patience and persistence during the editorial process, which was a tedious and time-consuming task. They did an excellent job cooperating with the editors in a timely manner.

All chapters in the book have received extensive examination by a review committee. We are grateful to all the reviewers who contributed significantly in improving the quality of the book. Their contributions were not limited to critical comments on scientific content but also provided constructive and thoughtful suggestions. The discussions between authors and reviewers in some cases were intensive and intriguing. The review committee members are listed in the next pages. We highly appreciated their contributions and efforts to examine each chapter.

Our sincere and deep gratitude goes to many of our colleagues, associates and assistants for helping us in editing the book in time. In particular, Bing Zhou would like to thank his colleagues at Headwaters Technology Innovation Group, including Kelly Repoley, Rebecca Groenendaal, Patricia Livingstone, Jen Stone, Michael Elwell, Michael Rueter, Robert Stalzer, Rober Chang, and Sukesh Parasher. Sophie Hermans expresses her acknowledgements to Professor Michel Devillers for his support and patience throughout the realisation of this project, to Jacqueline Boniver for her secretarial assistance, and, most of all, to Benoît Poncin for his endless help and encouragement.

ACS Secretariat of Surface Science and Catalysis (CATL) was the principal sponsor of the symposia on Nanotechnology in Catalysis. We want to thank Dr. Nancy B. Jackson, Dr. Lisa S. Baugh, and Dr. David Bergbreiter, General Secretaries of CATL in 2000, 2001, and 2002 respectively, for their support and their encouragement to start and continue these symposia. Acknowledgements are also due to the ACS Divisions of Colloid, Petroleum, and Industrial & Engineering Chemistry for co-sponsoring the symposia. Financial support from the ACS Petroleum Research Fund and the Division of Colloid Chemistry is highly appreciated.

Pars Environmental Inc., a company located in Robbinsville, New Jersey also provided financial support for the first symposium. Bing Zhou would like to thank Dr.

Harch Gill, Chief Executive Officer, and Dr. Chiang Tai for their interest and involvement in promoting applications of nanocatalysts in environmental protection.

Review Committee

Richard T. Baker
University of Dundee
Dundee, United Kingdom

Dominique C. Bazin
Paris XI University
Orsay, France

Jingguang G. Chen
University of Delaware
Newark, Delaware

Paul A. Christensen
University of Newcastle upon Tyne
Newcastle, United Kingdom

Thomas F. Degnan, Jr.
Exxon Mobil Research & Engineering Company
Annandale, New Jersey

Krijn P. de Jong
Utrecht University
Utrecht, The Netherlands

Junfeng Geng
University of Cambridge
Cambridge, United Kingdom

Malcolm L.H. Green
University of Oxford
Oxford, United Kingdom

Ting Guo
University of California
Davis, California

James F. Haw
University of Southern California
Los Angeles, California

Sophie Hermans
Catholic University of Louvain
Louvain-la-Neuve, Belgium

Challa S. S. R. Kumar
Lousiana State University
Baton Rouge, Louisiana

Rajiv Kumar
National Chemical Laboratory
Pune, India

Jingyue Liu
Monsanto Company
St. Louis, Missouri

Charles M. Lukehart
Vanderbilt University
Nashville, Tennessee

Bill McCarroll
Rider University
Lawrenceville, New Jersey

Chung-yuan Mou
National Taiwan University
Taipei, Taiwan

Martin Muhler
der Ruhr-Universität Bochum
Bochum, Germany

Janos B. Nagy
Facultes Universitaires Notre-Dame de la Paix
Namur, Belgium

Radha Narayanan
Georgia Institute of Technology
Atlanta, Georgia

Sang-Eon Park
Korea Research Institute of Chemical Technology
Taejon, Korea

Gabor A. Somorjai
University of California
Berkeley, California

James C. Vartuli
ExxonMobil Research & Engineering Company
Annandale, New Jersey

Karen Wilson
University of York
York, United Kingdom

Chuan-Jian Zhong
State University of New York
Binghamton, New York

Bing Zhou
Headwaters NanoKinetix, Inc.
Lawrenceville, New Jersey

Contents
Volume I

Section I
Fundamental Understanding of Catalysis at the Nanoscale

CHAPTER 1.   Catalysis and Nanoscience

J. Grunes, J. Zhu, and G. A. Somorjai

Section II
Synthesis, Characterization and Reaction Study of Nanoparticle
and Nanocluster Catalysts

CHAPTER 2.   Novel Catalytic Properties of Bimetallic Surface Nanostructures

N.A. Khan, J.R. Kitchin, V. Schwartz, L.E. Murillo, K.M. Bulanin, and J.G. Chen

CHAPTER 3.   Molecular Mixed-Metal Clusters as Precursors for Highly Active
Supported Bimetallic Nanoparticles

CHAPTER 6.   Gold Nanoparticles Formed within Ordered Mesoporous Silica and on Amorphous Silica

R. Kumar, A. Ghosh, C. R. Patra, P. Mukherjee, and M. Sastry

**CHAPTER 7.   Multifunctional Catalysts for Singlewall Carbon Nanotube
                Synthesis**

T. Guo

**CHAPTER 8.   Nickel and Ruthenium Nanoparticles as Catalysts for Growth of
                Carbon Nanotubes and Nanohorns**

J. Geng, and B. F. G. Johnson

CHAPTER 9.   A Novel Route to Prepare Pt-Based Electrocatalysts for Direct Methanol (Ethanol) Fuel Cells

W. J. Zhou, B. Zhou, Z. H. Zhou, W. Z. Li, S. Q. Song, Z. Wei, G. Q. Sun, and Q. Xin

CHAPTER 10.   Pt-Ru/Carbon Nanocomposites: Synthesis, Characterization and Performance as DMFC Anode Catalysts

W. D. King, E. S. Steigerwalt, G. A. Deluga, J. D. Corn, D. L. Boxall, J. T. Moore, D. Chu, R. Jiang, E. A. Kenik, and C. M. Lukehart

CHAPTER 11.   Nanostructured Gold and Alloy Electrocatalysts

C. J. Zhong, J. Luo, M. M. Maye, L. Han, and N. Kariuki

Section III
Ordered Nanoporous Materials as Catalysts or Support

**CHAPTER 12.   Shape-Selective Regiospecific and Bifunctional Nanoporous
Catalysts for Single Step Solvent Free Processes**

R. Raja, and J.M. Thomas

CHAPTER 13.   Nanofunctionalized Microporous Catalysts

J. F. Haw, and D. M. Marcus

CHAPTER 14. New Catalytic Materials for Clean Technology: Structure-Reactivity Relationships in Mesoporous Solid Acid Catalysts

K. Wilson, A. F. Lee, M. A. Ecormier, D. J. Macquarrie, and J. H. Clark

CHAPTER 15.   Mesoporous Materials as Catalyst Supports

C. L. Chen, and C. Y. Mou

CHAPTER 16. Microwave-Induced Synthesis and Fabrication of Nanoporous Materials

S.-E. Park, Y. K. Hwang, D. S. Kim, J.-S. Chang, J. S. Hwang, and S. H. Jhung

CHAPTER 17. **Nanosized β Zeolites and Their Composites as Catalysts for
Acylation and Alkane Isomerization**

Y. R. Wang, E. Z. Min, and X. H. Mu

Contents

Volume II

**CHAPTER 19. Nano-Scale Characterisation of Supported Phases in Catalytic
Materials by High Resolution Transmission Electron Microscopy**

R. T. Baker, S. Bernal, J. J. Calvino, J. A. Pérez-Omil, and C. López-Cartes

**CHAPTER 20. Solid State Physics and Synchrotron Radiation Techniques to
Understand Heterogeneous Catalysis**

D. C. Bazin

CHAPTER 21. **Design, Building, Characterization and Performance at the Nanometric Scale of Heterogeneous Catalysts**

J.-M. Basset, J.-P. Candy, C. Copéret, F. Lefebvre, and E. A. Quadrelli

CHAPTER 22. **Modelling Transition Metal Nanoparticles: the Role of Size Reduction in Electronic Structure and Catalysis**

L. Guczi, Z. Pászti, and G. Pető

Section V
New Nanomaterials and Applications in Catalysis

CHAPTER 23. Nanoscale Dendrimer-Supported Hydrofomylation Catalysts for Membrane Separations

M. L. Tulchinsky, and J. C. Hatfield

CHAPTER 24. The Catalytic/Oxidative Effects of Iron Oxide Nanoparticles on Carbon Monoxide and the Pyrolytic Products of Biomass Model Compounds

P. Li, E. J. Shin, D. E. Miser, F. Rasouli, and M. R. Hajaligol

CHAPTER 25. Graphitic Nanofilaments: A Superior Support of Ru-Ba Catalyst for Ammonia Synthesis

C. H. Liang, Z. L. Li, J. S. Qiu, Z. B. Wei, Q. Xin, and C. Li

Nanotechnology in Catalysis
Volume 2

18

Advanced Electron Microscopy in Developing Nanostructured Heterogeneous Catalysts

J. Liu*

18.1. INTRODUCTION

Catalysis is the science of chemical reactions on an atomic or molecular level. Using the modern language, we can define catalysis as a process of facilitated self-assembly of atoms or molecules; and a catalyst acts as the facilitator. Heterogeneous catalysis refers to chemical reactions that occur on a surface; it involves solid catalysts and gaseous or liquid reactants. Heterogeneous catalysts used in commercial processes are often chemically and physically complex systems that have been developed through many years of catalytic art, science, and technology. Industrial catalysis still defies our understanding because the catalyst preparation methods practiced commercially are not well controlled and the synthesis-structure-performance relationships are poorly understood.

Nanoclusters and nanoparticles are key components of supported catalysts, which is one of the most important classes of heterogeneous catalysts. The use of nanoparticles is not just for significantly increasing the total surface area of a catalyst; nanoparticles also possess unique electronic, magnetic, optical, and catalytic properties. Kinks, steps, corners, and different surface planes on a nanoparticle provide various specific sites that can bind and/or dissociate reactant molecules. The reactivity of a nanoparticle depends on its size, shape, surface composition, and surface atomic arrangement. Each

* Jingyue Liu, Science & Technology, Monsanto Company, 800 North Lindbergh Boulevard, U1E, St. Louis, Missouri 63167, USA. Email: jingyue.liu@monsanto.com

nanocluster with a specific number of atoms has its own intrinsic chemical and physical properties and the size-dependent quantum effects can become dominant.

To understand the behavior of nanostructured heterogeneous catalysts, we need to redefine how we describe their physical characteristics. Some concepts commonly used to describe the nature of heterogeneous catalysts are ambiguous at best and misleading in some cases. For example, the definition of dispersion (D) as the fraction of atoms of an active phase exposed to the surface has unique meaning when we apply it to a single nanoparticle. But when we apply the same definition to describe a heterogeneous catalyst that consists of individual nanoparticles with different sizes, the meaning of "dispersion" is ambiguous or misleading. The same argument applies to the concept of the "average size" of nanoparticles. Since the performance of a heterogeneous catalyst critically depends on the size distribution, rather than the average size, of the active nanoparticles, we should use the size distribution, the surface area distribution, or the volume distribution of the individual nanoparticles to describe the physical nature of heterogeneous catalysts. We should also develop methods to quantitatively, or at least semi-quantitatively, describe the shape and spatial distribution of individual nanoparticles since these parameters can strongly influence the catalytic performances of heterogeneous catalysts. Size- and shape-controlled synthesis of nanoparticles is critical to developing nanostructured catalysts.

Catalyst characterization is the cornerstone of the science of catalysis. Not only we need to know the electronic and geometric structure of the active nanocomponents in a heterogeneous catalyst but also we need to understand how the reactant molecules interact with or modify these nanocomponents during a catalytic reaction. Catalyst characterization and development are most challenging and are at the forefront of materials science, chemistry, surface science, chemical engineering, and physics.

Supported catalysts are widely used commercially and generally consist of complex nanostructures: the pore structure of the support is non-uniform; the size and shape of the nanoparticles vary from particle to particle; the interfacial structure between the nanoparticles and the support is complex and is poorly understood; and the surface composition and structure of the individual nanoparticles may depend on their sizes, relative locations, and methods of preparation and treatment. Generally speaking, however, a simple "raisin in the sponge cake" model may conceptually describe a well-dispersed, supported metal catalyst. Just to illustrate this point, Fig. 18.1a shows an idealized schematic model of the structure of metal nanoparticles supported on high-surface-area γ-alumina crystallites. The pores within large agglomerates of γ-alumina crystallites are formed by randomly stacking individual γ-alumina nanoplatelets or nanoneedles. The metal nanoparticles generally have different shapes and sizes; their interaction with the γ-alumina crystallites can be complex and may depend on the location, size, and shape of the individual nanoparticles as well as the conditions and history of the catalyst preparation and treatment. The use of interconnected γ-alumina crystallites with large aspect ratios provides high-surface area to disperse the metal nanoparticles and also provides highly tortuous pathways to prevent migration and agglomeration of metal nanoparticles during the catalyst preparation processes and during the catalytic chemical reactions. Figure 18.1b is a high-angle annular dark-field (HAADF) image of an industrial $Pd/\gamma-Al_2O_3$ catalyst; it clearly shows numerous interconnected pores of various sizes and the high tortuosity as well as the non-uniform size distribution of the palladium nanoparticles (bright dots in the image). The accessibility of the palladium nanoparticles by the reactant molecules is determined by the relative locations of the nanoparticles with respect to the interconnected pores of the alumina support. Therefore, not only the size distribution but also the spatial distribution of metal nanoparticles is critical to understanding the performance of supported metal catalysts. In practice, each commercial catalyst has its own unique physicochemical properties; and often these properties are difficult to be controlled or characterized.

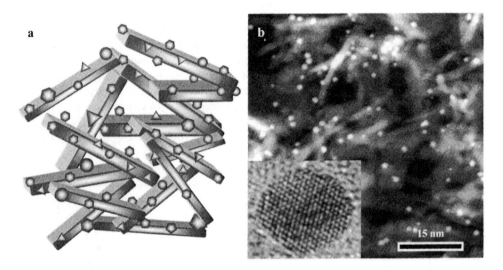

FIGURE 18.1. A schematic model of a γ-Al₂O₃ supported Pd catalyst (a) and an experimental high-angle annular dark-field image of an industrial Pd/ γ-Al₂O₃ catalyst showing Pd nanoparticles and the γ-Al₂O₃ nanoplatelets (b). The inset in (b) shows an atomic resolution image of a Pd nanoparticle (Reprinted with the permission of the Cambridge University Press and the Microscopy Society of America).[20]

The next level of characterization of the catalyst shown in Fig. 18.1b is to examine the shape, arrangement of surface atoms, and the electronic structure of the individual Pd nanoparticles. The inset in Fig. 18.1b is a high-resolution electron microscopy (HREM) image, showing the atomic structure and the shape of the Pd nanoparticle. The binding energy of the Pd surface atoms can vary at different sites of the Pd nanoparticle.

The interfaces between the Pd nanoparticles and the γ-Al₂O₃ nanoplatelets should also be characterized, understood, and preferably controlled. A complete characterization of industrial catalysts requires extensive resources and is often extremely expensive and slow because of the complex nature of the industrial catalysts; this is why we still do not understand the behavior of most of the heterogeneous catalysts used commercially.

Among many characterization techniques, advanced electron microscopy techniques are the only techniques that can provide information on the individual nanocomponents of industrial heterogeneous catalysts. All the other techniques (e.g., X-ray diffraction, X-ray absorption spectroscopy, infrared spectroscopy, nuclear magnetic resonance spectroscopy, X-ray photoelectron spectroscopy, etc.) provide information either averaged over millions to trillions of nanocomponents or they (e.g., scanning tunneling microscopy and atomic force microscopy techniques) require stringent conditions on the samples to be examined. Therefore, advanced electron microscopy techniques are indispensable for understanding the properties of heterogeneous catalysts and for guiding the development of nanostructured industrial catalysts.

18.2. ELECTRON MICROSCOPY IN DEVELOPING NANOSTRUCTURED CATALYSTS: AN OVERVIEW

Because of the recent explosive research activities in nanoscience and nanotechnology advanced electron microscopy techniques have been becoming a more and more powerful tool for characterizing nanophase materials, nanodevices, nanoparticles, nanostructured catalysts, and other nanosystems. To illustrate this point, Fig. 18.2 shows the result of a quick analysis of the number of published papers each year that used one or more of the electron microscopy techniques to characterize catalysts or nanoparticles. The graphs clearly demonstrate that electron microscopy techniques are becoming a useful tool to understand the nanoscale properties of heterogeneous catalysts.

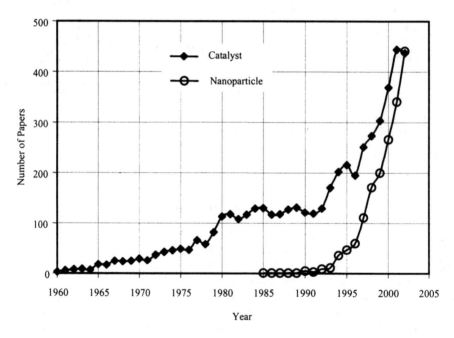

FIGURE 18.2. Histograms show the number of papers published each year that uses one or more of electron microscopy techniques to characterize catalysts or nanoparticles.

The rapid adoption of electron microscopy techniques by the catalysis community in the last few years coincides with the rapid expansion of research in nanoscience; for example, electron microscopy techniques are clearly widely used for characterizing nanoparticle systems. Advanced electron microscopy techniques, especially atomic resolution electron microscopy and spectroscopy techniques, are indispensable for studying the fundamental properties of nanoparticles and for guiding the synthesis of nanoparticles and nanostructured catalysts with desired structural properties.

The one most single important feature of a modern electron microscope is its versatility: atomic resolution images, diffraction patterns from nanometer regions, and nanometer-scale spectroscopy data can be obtained either simultaneously or sequentially from the same region of the specimen. The availability of the various imaging, diffraction, and spectroscopy techniques within a single instrument makes the modern electron microscope a powerful tool for characterizing the physicochemical nature of nanoscale systems, especially nanostructured heterogeneous catalysts.

When an electron beam interacts with a thin specimen inside an electron microscope, a variety of electron, photon, phonon, and other signals can be generated. Figure 18.3 shows a schematic diagram illustrating the common signals that are produced by strong electron-solid interactions. There are three types of transmitted electrons: 1) non-scattered electrons, 2) elastically scattered electrons, and 3) inelastically scattered electrons. There are also three types of electrons that can be emitted from the electron-entrance surface of the specimen: 1) secondary electrons with energies < 50 eV, 2) Auger electrons produced by the decay of the excited atoms, and 3) backscattered electrons that have energies close to those of the incident electrons.

All these signals can be used to form images or diffraction patterns of the specimen or can be analyzed to provide spectroscopic information. For example, electron energy-loss spectroscopy (EELS), which is based on the energy analysis of the inelastically scattered electrons, can provide information on the electronic structure, oxidation states, and chemical composition of catalysts on an atomic or sub-nanometer scale.

The de-excitation of atoms that are excited by the primary electrons also produces continuous and characteristic X-rays as well as visible light. These signals can be utilized to provide qualitative and quantitative information on the elements present in the regions of interest. X-ray energy dispersive spectroscopy (XEDS), for example, is a powerful tool for identifying elements that are present in a heterogeneous catalyst. We can transform XEDS spectra into quantitative data describing different phases and phase distributions in heterogeneous catalysts.

Instead of discussing the basic principles of specific electron microscopy techniques, we will review, in this chapter, the most recent developments in advanced electron microscopy and how to apply these new techniques to the study of nanostructured catalysts. We will also emphasize on an integrated microscopy approach to solving practical problems encountered in developing industrial catalysts. Those readers who are interested in learning more about the specific electron microscopy techniques should consult the following textbooks or review articles: high-resolution electron microscopy[1, 2]; transmission electron microscopy techniques[3]; electron energy-loss spectroscopy[4]; X-ray energy dispersive spectroscopy[5]; scanning transmission electron microscopy techniques[6-11]; and scanning electron microscopy techniques.[12]

Since several review articles on the applications of conventional and high-resolution electron microscopy to characterizing heterogeneous catalysts were published in the early 1990s[13-15], we will primarily focus here on discussing the recent developments of advanced electron microscopy techniques and demonstrating that these techniques are indispensable for developing nanostructured catalysts. Furthermore, since the chapter immediately following this chapter discusses in detail the simulation of high-resolution transmission electron microscopy images and its applications to the study of supported metal catalysts, we will emphasize, in this chapter, on discussing high-resolution analytical electron microscopy techniques, surface electron microscopy techniques, and other special electron microscopy techniques that are relevant to characterizing heterogeneous nanostructured catalysts.

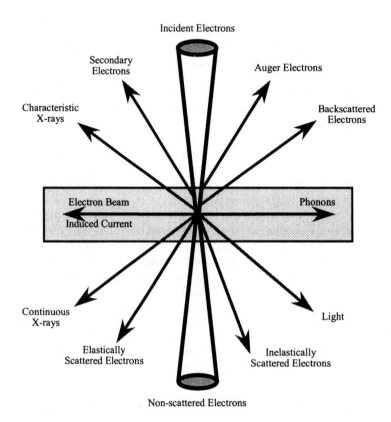

FIGURE 18.3. Schematic diagram illustrates the available signals generated inside an electron microscope due to the strong interaction of high-energy electron beam with a thin specimen. All these signals can be utilized to form images, diffraction patterns, or spectra that carry information on the nature of the specimen (Reprinted with the permission of WILEY-VCH).[11]

18.3. ATOMIC RESOLUTION TRANSMISSION ELECTRON MICROSCOPY OF HETEROGENEOUS CATALYSTS

Atomic resolution imaging can now be routinely achieved in the modern transmission electron microscope (TEM) and has been widely applied to characterizing the atomic structure of a plethora of materials.[1] Although scanning tunneling microscopy and atomic force microscopy techniques can also provide atomic resolution information and have proved invaluable for studying the fundamental processes of heterogeneous catalysis,[16] these techniques have not been successfully applied to the study of practical industrial catalysts. On the other hand, most heterogeneous catalysts can be directly examined in an electron microscope.

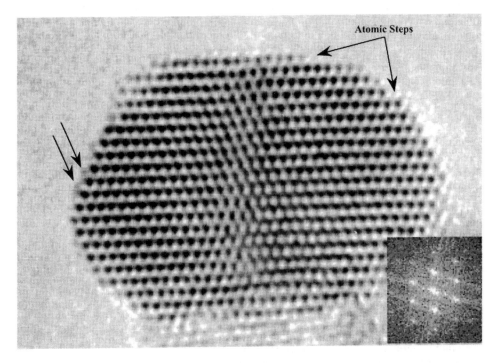

FIGURE 18.4. HREM image of an antimony-doped SnO$_2$ nanoparticle shows surface steps, facets, restructuring of surface atoms, and an internal defect. The inset shows the corresponding digital diffractogram clearly revealing that the nanocrystal has a rutile-type structure.

Atomic structure of heterogeneous catalysts such as facets, surface steps, and surface reconstructions can be visualized by using a technique called surface profile imaging;[1, 17] in the profile-imaging mode, a thin crystal is aligned such that the direction of the electron beam is parallel to a low-index zone-axis of the crystal. As an example, Fig. 18.4 shows an atomic resolution profile image of a small crystal of an antimony-doped tin oxide (ATO) nanocatalyst. The ATO nanocatalyst is of interest in the oxidation of propylene to acrolein and the selective oxidation of olefins. The image was recorded at the optimum defocus value so that the atomic columns appear as black dots in the image.[1] Surface steps, rearrangement of surface atoms (indicated by the double arrows), facets, and an internal defect, which is probably caused by the incorporation of Sb dopants, are clearly revealed. The inset shows the corresponding digital diffractogram clearly revealing that the Sb-doped SnO$_2$ nanocrystallite has a rutile-type structure. The effect of the antimony doping and the calcination temperature on the growth of SnO$_2$ nanocrystallites as well as the oxidation states of the Sb dopant has recently been studied by the combined use of HREM, HAADF, and EELS techniques.[18]

Atomic structure and structural evolution of many heterogeneous catalysts can be examined by the profile imaging method. Surface reconstructions of small gold particles and adsorbate-induced restructuring of Pt metal surfaces were investigated by the HREM technique.[17, 19] HREM examination of fresh and used catalysts can often yield information on the surface modification of metallic nanoparticles as well as their sintering behavior during catalytic reactions.

In supported metal catalysts, the performance of a catalyst can be directly related to the available number of active sites of the metal nanoparticles; thus, it is important to determine their sizes and size distributions. For nanoparticles larger than about 3 nm in diameter, XRD techniques, especially when synchrotron sources are used, can provide the required information; however, the sizes provided by the XRD technique reflects the sizes of the single crystallites which may not be the sizes of the nanoparticles. For nanoparticles less than 3 nm in diameter, it is difficult, if not impossible, to obtain accurate information about the size distributions of the metal nanoparticles by the XRD technique. On the other hand, small particles, clusters, or even single atoms can be directly observed in the modern electron microscope. In the following, we will discuss the particle visibility, structure and shape of metal nanoparticles, the measurement of lattice parameters of small nanoparticles, and the limitations of applying the HREM techniques to the study of nanoparticles and supported metal catalysts.

18.3.1. Visibility of Nanoparticles in Supported Catalysts

18.3.1.1. *Plan-View Imaging.* The visibility of metal nanoparticles supported on light-element carriers depends on the atomic number of the metal particle, the shape of the particle, and the ratio of the size of the metal particle to the thickness of the support. For example, when supported on extremely thin carbon films, clusters of platinum atoms or even individual Pt atoms can be directly imaged.[20]

When the thickness of the support increases, however, the visibility of very small metal clusters drastically decreases; details of the shape and structure of the nanoparticles are then often overshadowed by the dominant phase contrast of the support and sub-nanometer particles can become completely invisible. When the support material is amorphous, a so-called "minimum contrast" technique can be applied to enhance the visibility of metal nanoparticles.[21] In general, however, when imaging metal nanoparticles with a size less than 3 nm in diameter one has to be cautious in interpreting the HREM images.[22, 23] For observing industrial catalysts, however, small metal nanoparticles are only visible near the edges of the support powders.

When the support material is crystalline in nature, the dominant fringe contrast from the support can overshadow or severely interfere the contrast of supported nanoparticles. The optimum-focus (also called the extended Scherzer focus) condition commonly used in HREM imaging[1] may not be the most favorable imaging condition for detecting small metal nanoparticles supported on crystalline materials. The particle visibility generally increases with the increase in defocus value. The variations in the image contrast are related to the dependence of the lens transfer function on the defocus value: at certain spatial frequencies the relative contrast of the metal nanoparticles is enhanced while the phase contrast of the crystalline support is suppressed. To illustrate this point, Fig. 18.5 shows HREM images of the same region of a Pd/TiO$_2$ catalyst but obtained at different imaging conditions. The defocus value for Fig. 18.5a was optimized for enhancing the fringe visibility of the TiO$_2$ crystallite; there were no observable features that represent Pd nanoparticles.   On the other hand, Fig. 18.5b, which was obtained by further defocusing the electron beam, clearly shows the presence of nanometer or sub-nanometer Pd clusters. This example demonstrates that the visibility of small nanoparticles strongly depends on the imaging conditions of the electron microscope.   Quantitative interpretation of HREM images of small nanoparticles in supported catalysts, however, requires detailed image simulations.[24]

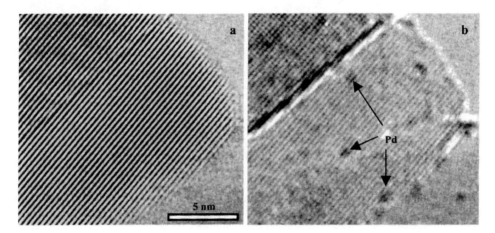

FIGURE 18.5. HREM images of the same region of a Pd/TiO$_2$ catalyst show the dependence of the visibility of Pd nanoparticles on the imaging condition of the electron microscope (see text for details) (Reprinted with the permission of the Cambridge University Press and the Microscopy Society of America).[20]

18.3.1.2. Profile-View Imaging.   In profile-view imaging of supported catalysts, the electron beam travels parallel to the contact plane between a metal particle and the support; thus the interface between the metal nanoparticle and the support can be imaged in cross section. Even if the electron beam is not parallel to the contact plane, which is usually the case when examining industrial catalysts, detailed information about the surface atomic arrangement of a metal nanoparticle can still be obtained if at least part of the nanoparticle is hung over vacuum. This imaging mode has proved to be invaluable for studying surface structures of a variety of materials.[1]

The profile-imaging technique is not generally applicable for examining commercial heterogeneous catalysts since only a very small percentage of the metal nanoparticles are attached to the edges of support powders. This technique, however, has been extensively used to study the surface structure of supported metal nanoparticles and the strong metal-support interactions.[14, 25]   In fact, one of the most significant contributions of HREM to understanding the nature of supported catalysts is the direct experimental evidence, revealed in the HREM images, of the decoration and encapsulation (geometric effect) of metal nanoparticles by the support material.[26-28]   For example, the decoration of the Pd nanoparticle by the support material is clearly revealed in the Fig. 18.6 of a HREM image of a Pd/TiO$_2$ catalyst (indicated by the white arrows); TiOx species may have migrated onto the surfaces of the Pd nanoparticles during the reduction processes.

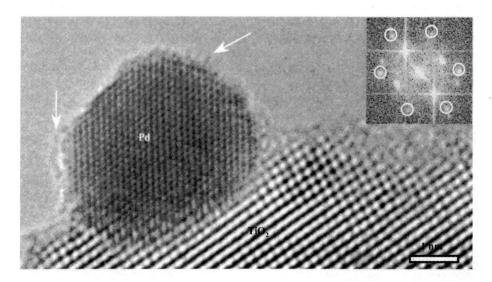

Figure 18.6. HREM image of a Pd/TiO₂ catalyst shows the surface decoration of the Pd nanoparticle by a layer of TiOₓ material. The inset is the corresponding digital diffractogram of the HREM image. The white dots within the circles represent the Pd (111) and (200) diffraction spots (Reprinted with the permission of the Cambridge University Press and the Microscopy Society of America).[20]

Detailed image simulations and quantitative analysis of HREM images of supported catalyst nanoparticles will be discussed in the next chapter. The effect of particle size, the support thickness, the particle orientation, and the relative particle position with respect to the electron entrance surface of the support on the contrast of metal nanoparticles has been extensively studied.[29]

18.3.2. Structure and Shape of Metal Nanoparticles

The shapes of nanoparticles play an important role in determining the performance of supported metal catalysts. In structure-sensitive catalytic reactions, for example, only certain surface planes provide dominant active sites for the desired products; other surface planes may be either not active or active for producing undesired side products. The particle shape and morphology determine the relative amount of edge and corner atoms. For example, nanoparticles with a cubic shape have only {100} faces exposed while decahedral or icosahedral nanoparticles have only {111} faces exposed. If truncated octahedral nanoparticles are formed, both {111} and {100} faces are exposed to the reactants. Shape-controlled synthesis of supported metal nanoparticles and the characterization of these nanoparticles are of practical importance for developing nanostructured catalysts.

HREM images can provide information on the shape and, in some cases, the surface atomic structure of the nanoparticles. Internal structural defects such as twins can also be revealed; dislocations and stacking faults are rare in small metal nanoparticles although they do exist in large particles. Figure 18.7a shows an image of a gold decahedral particle oriented near the five-fold-axis orientation. Figure 18.7b is the corresponding digital diffractogram, clearly showing the ten (111) and ten (200) diffraction spots, reflecting the five-fold symmetry of decahedral particles. Many atomic steps can be identified along the edges of the decahedral particle and both {111} and {100} facets were formed. The insets in Fig. 18.7a show the corresponding digital diffractograms obtained from each individual tetrahedral domains of the decahedral nanoparticle projected along the five-fold axis. These diffractograms are not identical, however, and some of them are severely distorted from the normal FCC (face centered cubic) structure. The causes of these distortions may reflect the true structural distortions of the individual tetrahedron or may originate from the imaging artifacts.[30-31]

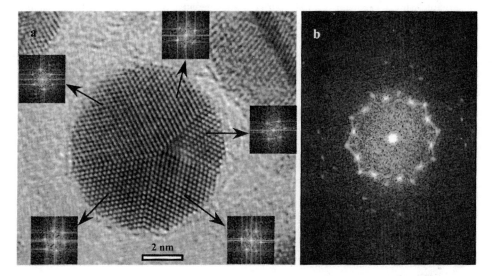

FIGURE 18.7. HREM image of a gold decahedral nanoparticle (a) and the corresponding digital diffractogram (b) show the five-fold symmetry of the nanoparticle. The insets in (a) show the digital diffractograms obtained from each individual tetrahedral domains of the decahedral nanoparticle (Reprinted with the permission of the Cambridge University Press and the Microscopy Society of America).[20]

The shapes of metal nanoparticles in supported metal catalysts depend on many parameters including their sizes, the method of preparation, the history of the catalyst treatment, and their interactions with the support. Even for model nanoparticles, it is difficult, if not impossible, to obtain statistically meaningful results on the shape distributions of nanoparticles. Therefore, we have not been able to meaningfully correlate the shapes of metal nanoparticles to their catalytic properties. Extensive research, however, has been conducted on the fundamental understanding of the structure and shape of metal nanoparticles and their formation mechanisms, with a focus on those nanoparticles that have a five-fold symmetry.[32-34] Experimental HREM images of metal nanoparticles that show five-fold-symmetry and image simulations of the observed particle contrast have been extensively investigated.[35, 36]

18.3.3. Measurement of Lattice Parameters

HREM technique is now well established for accurately measuring the lattice parameters, distortions of atomic arrangements near defect sites and interfaces, and surface reconstructions.[1] For supported metal catalysts, however, the metal nanoparticles are usually of irregular shape and are randomly oriented. In the reciprocal space, small particles are associated with large diffraction volumes at each reciprocal lattice point (the so-called shape effect); the consequence of this large diffraction volume is that even if a nanoparticle is tilted away from its exact zone axis, lattice fringes can still be observable. This is why most of the small particles show lattice fringes in HREM images even if the orientation of the nanoparticles is randomly distributed. For FCC metal particles, the observable fringes are usually {111} lattice planes because of their larger lattice spacings. Image simulations showed that simple interpretation of HREM images of randomly oriented nanoparticles be problematic because the observed fringe spacings may not correspond to the true lattice spacings of interest.[30, 31, 37] The variations in the measured fringe spacings can be as large as 12% from the true lattice spacings, especially for very small clusters and for fringes near the edges of a nanoparticle.[31]

Not only the spacings of the lattice fringes of randomly oriented nanoparticles can be highly distorted but also the angles between different sets of lattice fringes can be drastically altered.[31] To correctly determine the lattice spacings of small nanoparticles in HREM images a practical protocol and a statistical method such as the one proposed by Tsen et al. has to be used.[31]

18.4. ATOMIC RESOLUTION SCANNING TRANSMISSION ELECTRON MICROSCOPY AND ANALYTICAL ELECTRON MICROSCOPY OF SUPPORTED HETEROGENEOUS CATALYSTS

18.4.1. Introduction

Scanning transmission electron microscopes (STEM) are tailor-made to provide structural, chemical, and morphological information on heterogeneous catalysts on a nanometer or atomic level. A remarkable capability of a STEM instrument is the formation of a high-brightness electron probe with diameters < 0.2 nm at 100 keV and as small as 0.13 nm at 300 keV. To form such an electron nanoprobe, the use of a field-emission gun (FEG) is necessary to provide sufficient signal strength for viewing and recording STEM images, spectra, and diffraction patterns. Another attractive feature of STEM is its great flexibility in the detection system. Different signals (see Fig. 18.3 and Fig. 18.8) generated from the same sample area can be collected either independently or simultaneously and can be analyzed in parallel to yield complementary information.

Unlike in TEM, where a stationary, parallel electron beam is used to form images, the STEM is a mapping device. In a STEM instrument, a fine electron probe, formed by using a strong objective lens to de-magnify a small FEG electron source, is scanned over a specimen in a two-dimensional raster. Signals generated from the specimen are detected, amplified, and used to modulate the brightness of a second electron beam that is scanned synchronously with the STEM electron probe across a cathode-ray-tube (CRT) display. Therefore, a specimen image is mapped onto the CRT display for observation or is digitally stored in a computer. The resolution of STEM images is primarily determined by the incident probe size, the stability of the microscope, and the inherent properties of the signal generation processes.

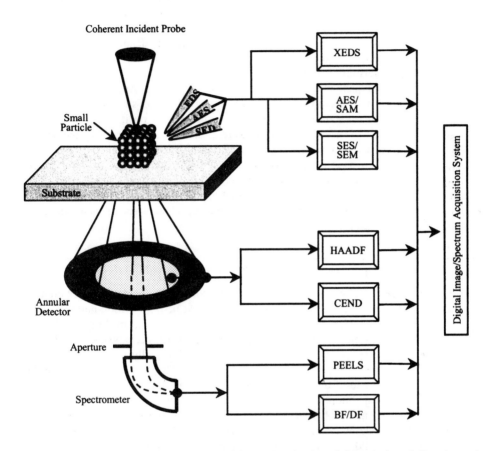

FIGURE 18.8. Schematic diagram illustrates the available signals and detectors in a dedicated scanning transmission electron microscope. XEDS---X-ray energy dispersive spectroscopy; AES---Auger electron spectroscopy; SAM---scanning Auger microscopy; SES---secondary electron spectroscopy; SEM---secondary electron microscopy; HAADF---high-angle annular dark-field; CEND---coherent electron nanodiffraction; PEELS---parallel electron energy-loss spectroscopy; BF---bright field; DF---dark field (Reprinted with the permission of WILEY-VCH).[11]

Multiple imaging and analytical detectors have been developed to simultaneously collect several signals that can be displayed individually or combined in perfect register with each other. This unique capability makes the STEM a powerful nanoanalytical tool since multiple views of a sample, in imaging, diffraction, and spectroscopy modes, can be collected, analyzed, and compared in a single pass of the incident electron beam. Figure

18.8 illustrates the available detectors commonly used for collecting imaging and analytical signals in high-resolution STEM instruments.

By collecting high-angle scattered electrons with an annular detector, HAADF images (or Z-contrast images) can be formed to provide information about structural variations across the sample at an atomic level. EELS and XEDS can give quantitative data describing changes of elemental composition, electronic structure, or state of oxidation associated with inhomogeneous structures of the sample. The combination of XEDS and EELS with HAADF imaging technique can provide detailed information on the composition, chemistry, and electronic and crystal structure of heterogeneous catalysts with atomic resolution and high sensitivity.

By collecting and analyzing secondary electron (SE) and Auger electron (AE) signals we can extract information on surface topography and surface chemistry of heterogeneous catalysts with nanometer spatial resolution. By positioning an electron nanoprobe at the area of interest, diffraction patterns from nanometer or subnanometer regions can be acquired to provide local crystallographic structure of nanoparticles. The structural relationship of nanoparticles to the surrounding materials and, in some cases, the shape of nanoparticles, can also be extracted from nanodiffraction patterns. The powerful combination of atomic resolution HAADF imaging with nano-spectroscopy and nano-diffraction techniques has proved invaluable in characterizing heterogeneous catalysts, especially supported metal and alloy catalysts.

18.4.2. High-Angle Annular Dark-Field Microscopy of Supported Catalysts

The high-angle annular dark-field microscopy technique is ideal for examining supported catalysts, especially those catalysts that consist of high-atomic-number metal particles dispersed onto light-element supports.[38-42] As shown in Fig. 18.1b, the diffraction and phase contrast, which are the dominant contrast mechanisms in TEM and HREM images, from the crystalline alumina support is drastically suppressed and the compositional sensitivity is recovered in HAADF images. The very high contrast of the Pd nanoparticles originates from the strong dependence of the strength of high-angle scattered electrons on the atomic number (Z) of the scatterers. In HAADF images, the signal strength of the high-angle scattered electrons is proportional to approximately Z^2. Therefore, HAADF imaging technique is ideal for detecting heavy-metal particles such as Bi, Au, Pt, Ir, Re, W, Pd, Rh, Ru, Ag, etc. supported on light-element carriers such as carbon, silica, alumina, magnesia, etc.

Since the electron probes in the modern electron microscopes can be made smaller than 0. 2 nm in diameter, clusters of atoms of heavy elements supported on light-element substrates can be visualized. Under optimum conditions, single atoms, dimers, trimers, and small nanoclusters can be observed.[42-44] With aberration-corrected STEM instruments, which we will discuss in section 18.6.2, sub-angstrom electron probes with high current can be obtained; single atoms, atomic monolayers, and clusters of atoms in supported catalysts can be imaged with high contrast.[45] The achievement of being able to form sub-angstrom electron probes opens tremendous opportunity for the study of the fundamental properties of heterogeneous catalysts.

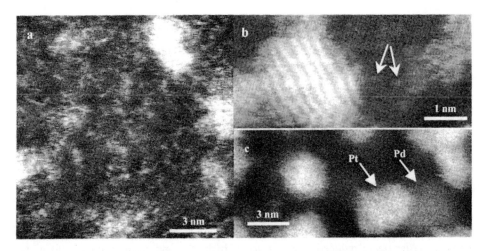

FIGURE 18.9. HAADF images of a carbon-supported Pt model catalyst show Pt single atoms or clusters (a) and the movement of Pt atoms to form islands (b). (c) HAADF image of a carbon-supported bimetallic model catalyst show the presence of individual Pt and Pd nanoparticles (Reprinted with the permission of Cambridge University Press and the Microscopy Society of America).[20]

Figure 18.9a shows a high-resolution HAADF image of a specially prepared model catalyst consisting of individual Pt atoms and nanoclusters of various sizes supported on a thin, amorphous carbon film, revealing various sizes of Pt nanoclusters; some of the bright spots may represent single Pt atoms or columns of Pt atoms. With the use of a hot stage, we have performed in situ experiments to examine the formation of small islands or clusters of Pt atoms (Fig. 18.9b). The image also provides information about the sintering behavior of the nanoclusters; for example, individual Pt atoms (indicated by the arrows) may move from one cluster to another, following the Ostwald-ripening mechanism. The sintering behavior of metal nanoclusters strongly depends on the metal-support interaction, the gas environment, and the reduction temperature. Fundamental understanding of the adsorption and adhesive energies of atoms and nanoclusters on various types of supports and how to anchor these nanoclusters can provide important information for controlling the long-term stability of nanostructured industrial catalysts.

The application of HAADF imaging to complex catalyst systems is desirable. For this purpose, we made a model catalyst consisting of separate Pt and Pd nanoparticles supported on a carbon film; Fig. 18.9c shows a HAADF image of such a system. By examining the intensity distribution, we can qualitatively differentiate Pt nanoparticles from Pd nanoparticles. For very small nanoclusters, however, it is difficult to differentiate Pt from Pd without quantitative image analysis, which has its own intrinsic problems. We found that the variations in the thickness of the support materials play a critical role in determining the wide applicability of HAADF imaging technique to characterizing mixed catalyst systems consisting of very small metal or alloy clusters. Again, with the use of Cs-corrected electron microscopes, we should be able to quantitatively analyze complex catalyst systems such as bimetallic and multimetallic catalysts or heterogeneous catalysts doped with various promoters.

18.4.3. High-Resolution Nanoanalysis by X-ray Energy-Dispersive Spectroscopy

XEDS is now routinely used, in TEM, SEM, or STEM instruments, to identify unknown phases or to obtain information about the spatial distribution of certain phases of interest. In a modern FEG TEM/STEM instrument, XEDS can be conveniently used to analyze the features revealed in HAADF images by stopping the incident probe at any point of interest. With the recent development of image and spectrum acquisition systems, both qualitative and quantitative information on the composition of individual nanocomponents can be obtained while the experiment is in progress. The rapid employment of faster computers for automation and online data analysis makes it possible now to analyze extremely complex industrial heterogeneous catalysts and to quickly diagnose the basic nature of the catalysts of interest.

Figure 18.10 shows a typical example of how to combine the techniques of HAADF and XEDS to investigate supported bimetallic catalysts. The sample is a γ-alumina supported Pd-Cu bimetallic catalyst with a nominal composition of 2wt%Pd and 1wt%Cu. After being reduced in a mixture of hydrogen and argon gases at 800° C for two hours, the nanoparticles were drastically sintered to form large particles (Fig. 18.10a). To understand the performance of bimetallic catalysts and to optimize the synthesis parameters, it is important to know how the Pd and Cu are mixed in the individual nanoparticles and how their composition varies with the particle size, the reduction temperature, and the reducing gases.[46] Figure 18.10b shows a XEDS spectrum obtained from the edge of the nanoparticle indicated by the arrow in Fig. 18.10a. The probe size used for obtaining the XEDS spectra was about 0.5 nm in diameter. By analyzing a series of XEDS spectra obtained from different regions of the nanoparticle, we concluded that the Pd and Cu were intimately mixed in this nanoparticle. Quantification of the composition of the bimetallic nanoparticles can then be obtained by using the Cliff-Lorimer equation.[3] Within the resolution limit (about 2 nm in this case), we did not detect any preferential surface aggregation of either Pd or Cu. The XEDS technique is, however, not sensitive to detecting variations in composition of the most outside surface layer of the alloy nanoparticles.

One of the most useful techniques for understanding the behavior of supported bimetallic catalysts and for guiding the development of industrial bimetallic catalysts is the composition-size plot method developed by Lyman's group.[47-49] The composition-size plots can provide the compositional profiles of the individual bimetallic nanoparticles or clusters; they reveal whether the compositions of individual nanoparticles vary with their sizes or with their relative locations with respect to the catalyst support. When ultramicrotomed samples are used, this method can quantitatively pinpoint how the compositional profiles vary with the supports, the catalyst treatment, or the catalyst preparation parameters. The composition-size plot method can also be applied to studying the compositional evolution of individual bimetallic nanoparticles during the catalytic reactions.

A consequence of using small electron probes to achieve high spatial resolution is that the X-ray signal originates from a much smaller volume; thus, a weaker signal is colleted and longer acquisition times are usually needed to obtain statistically meaningful results. Specimen-drift correction, either manually or automatically, is usually used for obtaining statistically meaningful X-ray signals when very small particles are analyzed. Nevertheless, XEDS can detect the presence of just a few atoms if the analyzed volume is small enough.[50]

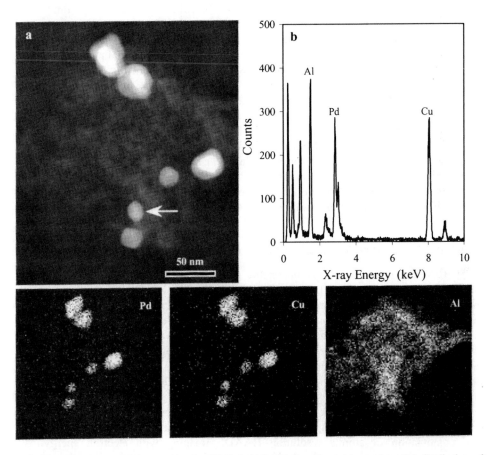

FIGURE 18.10. HAADF image (a) of a Pd-Cu/γ-Al₂O₃ catalyst shows size and spatial distribution of nanoparticles. XEDS spectrum (b) obtained from the nanoparticle (indicated in (a)) shows the presence of both Pd and Cu in the nanoparticle. X-ray elemental maps (Pd, Cu, and Al) of the catalyst are also shown (Reprinted with the permission of the Cambridge University Press and the Microscopy Society of America).[20]

With the recent development of Cs-correctors for dedicated STEM instruments, the total probe current can be significantly increased, thus providing higher counts of the generated X-ray signals. The effect of the intense electron beam on the structure of the catalyst may, however, set the limit of the usable probe current density. Future development that focuses on significantly improving the X-ray collection efficiency and high-throughput automatic analysis systems can have a profound impact on the fundamental understanding of heterogeneous catalysts.

Elemental maps provide valuable information on the two-dimensional elemental distributions in supported catalysts; they are especially useful for characterizing bimetallic or multiphase catalysts. Particles as small as 2 nm in diameter can be detected in digitally acquired X-ray maps.[51] Electron-specimen interaction and the volume of the

X-ray generation determine the ultimate resolution of X-ray mapping of nanoparticles. In practice, however, the extremely low counts of the collected X-ray signal limits the achievable resolution in X-ray maps. A statistically meaningful elemental map requires longer acquisition time that in turn requires the use of automatic drift correction or ultra-stable microscopes and samples. Figure 18.10 also shows the elemental maps of Pd, Cu, and Al of the same region as shown in Fig.18.10a. Correlation of the Pd and Cu maps suggests that within the image resolution ($\sim$ 2 nm) there is no preferential segregation of either Pd or Cu within the nanoparticles. Higher-resolution maps, however, are needed to determine the degree of surface segregations, which may be accomplished by using Cs correctors to reduce the probe size but still having enough beam current. Automatic specimen-drift correction may also have to be used to reduce sample-drifting effect. The conditions for optimum X-ray mapping include: 1) high beam current within a small probe size, 2) high X-ray collection efficiency, and 3) long acquisition times per pixel if automatic specimen-drift correction techniques are used. Instead of X-ray maps, line scans across features of interest are often used to extract the desired information.

18.4.4. High-Resolution Nanoanalysis by Electron Energy-Loss Spectroscopy

HREM can provide information on the atomic arrangement of nanoparticles and heterogeneous catalysts; HAADF imaging can provide the location of even very small metal clusters with high contrast, thus making the measurement of particle size distribution possible; XEDS can be used to identify the composition of these small clusters or nanoparticles. None of these techniques, however, can provide information about the oxidation states or the electronic structure of heterogeneous catalysts, which are most important for understanding the fundamental properties of heterogeneous catalysis. Furthermore, due to the intrinsic nature of the X-ray generation processes, XEDS cannot provide atomic resolution even if sub-angstrom electron probes are used. Electron energy-loss signals, on the other hand, carry detailed information about the composition, chemistry, and structure of supported catalysts with atomic resolution and sensitivity. The combination of atomic resolution HAADF imaging with electron energy-loss spectroscopy has already proved valuable for extracting atomic-scale information on the electronic structure of various materials.[52-55]

An unexplored area in supported metal catalysts is the effect of the electronic structure of the interfacial regions, between the metal nanoparticles and the support, on the catalytic performances of supported catalysts. Since catalytic reactions usually involve bonding and electron transfer processes, the electronic properties of the metal-support interfacial regions may play a critical role in determining surface adsorption and the electron transfer processes. Furthermore, the interfacial regions may also act as active sites during the catalytic reactions since these regions may have a structure that represents neither the metal nanoparticles nor the support. Knowledge of the atomic and electronic structure of the interfaces can help us better understand the performances of heterogeneous catalysts. Recently, we have applied the atomic resolution EELS and HAADF techniques to the study of nanophase $Sn(Sb)O_2$ catalysts,[18] metal-oxide interfacial structures in silica-supported Pt catalysts,[56] reduction behavior of metal nanoparticles in alumina-supported Pd catalysts,[57] and the alloying behavior of supported bimetallic catalysts.[46] We have also applied the atomic resolution EELS technique to the study of metal-support interactions and unexpected alloying phenomena in metal-oxide supported metal catalysts.[58] These preliminary investigations already showed that atomic resolution EELS together with HAADF imaging technique can provide new information on the fundamental understanding of the electronic structure as well as surface composition of the individual nanoparticles and their interactions with the support.

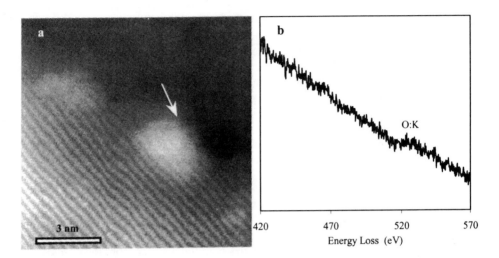

FIGURE 18.11. HAADF image of a Pt/TiO$_2$ catalyst (a) and the EELS spectrum (b) showing the presence of adsorbed oxygen on the Pt nanoparticle (Reprinted with the permission of the Cambridge University Press and the Microscopy Society of America).[20]

Not only the interfacial regions can be studied by atomic resolution EELS but also the surface electronic structure or adsorption properties of metal nanoparticles can be investigated. For example, Fig. 18.11a shows a high-resolution HAADF image of a titania-supported Pt catalyst, revealing the lattice fringes of the titania support as well as small Pt clusters. Part of the bigger Pt nanoparticle (~ 3 nm) shown in Fig. 18.11a (indicated by the arrow) was hung over vacuum. By positioning the 0.2 nm electron-probe at the edge of the particle an EELS spectrum was obtained as shown Fig. 18.11b. While there is no statistically meaningful signal at the Ti L-edge energy the oxygen K-edge peak is clearly observable. The oxygen signal shown in the EELS spectrum most probably originated from the oxygen that chemisorbed on the Pt surface or from an oxidized Pt surface layer. If the oxygen signal originated from the titania support or from TiO$_x$ species on the surface of the Pt nanoparticle, then a strong Ti peak would have been detected. Detailed analysis of this type of spectra can provide valuable information about the surface oxidation states and the adsorption properties of nanoparticles of different noble metals.

Similar to that of X-ray absorption spectroscopy techniques, energy-loss near-edge structure (ELNES) and extended energy-loss fine structure (EXELFS) techniques can be used to investigate the electronic structure of nanophase materials on a nanometer or sub-nanometer scale. As in the case of XANES (X-ray absorption near-edge structure) and EXAFS (extended X-ray absorption fine structure) studies, ELNES and EXELFS allow extracting information about the electronic structure and inter-atomic distances. The relatively low signal-to-noise ratio in the EELS spectra, however, limits the wide applications of EXELFS and ELNES techniques to characterizing nanophase materials and heterogeneous catalysts.

The power of applying atomic resolution EELS to the study of supported metal or alloy catalysts is further demonstrated in Fig. 18.12. The catalyst consisted of 2wt%Pd and 1wt%Ni dispersed onto nanophase titania support and was reduced at 800° C for two hours.[58] Figure 18.12a shows a HAADF image revealing a large particle supported on the TiO_2 powder. Both XEDS and EELS spectra obtained from the particle showed that it consists of intimately mixed Pd and Ni. Thus, Pd-Ni bimetallic alloy nanoparticles were formed during the high-temperature reduction process.

When the 0.2nm electron probe was placed at the edge of the bimetallic particle (indicated by the numeral 1 in Fig. 18.12a), however, the EELS spectrum (Fig. 18.12b) showed only the presence of the Pd peaks; the spectrum did not reveal any observable Ni signal. To verify the finding, more EELS spectra were obtained from the edge regions of the Pd-Ni nanoparticle and from regions just inside the surface of the particle. Two of these spectra are shown in Fig. 18.12c to show the Pd M-edges and in Fig. 18.12d to reveal the Ni L-edges. These spectra clearly demonstrate that the outmost surface layer of the Pd-Ni bimetallic nanoparticle consists of only Pd while the inner regions contain both Pd and Ni. Detailed analysis of many EELS spectra obtained from different regions of the nanoparticles can provide information about the thickness of the Pd skin layer and how the composition of the nanoparticles varies with the distance from the edges of the particles. This technique can be equally implemented in high-resolution environmental FEG TEM/STEM instruments to study the adsorbate-induced preferential surface segregation. Atomic resolution EELS technique practiced in the environmental TEM will provide vital information for understanding the physicochemical nature of bimetallic nanoparticles and on how the surface structure, composition, and morphology of individual nanoparticles respond to different gas environments.

The structure of the Pd-Ni bimetallic nanoparticles seems to follow that of the "grape" model: a thin skin of pure Pd layer encapsulates a Pd-Ni alloy core. This type of structure may have profound effect on the adsorption and catalytic properties of bimetallic catalysts. The knowledge of preferential surface segregation of individual bimetallic nanoparticles is critical to designing supported bimetallic catalysts and to understanding their performances. By knowing how the preferential surface segregation of individual bimetallic nanoparticles depends on the particle size, composition, and the catalyst preparation methods, we may be able to make significant progress in tuning the properties of supported bimetallic catalysts to achieve high selectivity and activity. The concept of binding-energy engineering refers to tuning the adsorption behavior of nanoparticles by controlling their sizes, atomic structure, shape, surface and bulk composition, and interface structure.

With the use of Cs-correctors, smaller electron probes with higher total beam current can be obtained.[45] The use of a monochromator in a FEG TEM/STEM can significantly increase the energy-resolution of the EELS spectra. When an energy resolution below 0.1 eV is achieved, many novel properties of individual nanoparticles can be explored. The combined use of a monochromator and a Cs-corrector in a modern FEG TEM/STEM microscope will undoubtedly provide us a powerful tool to probe the nature of nanoclusters, nanoparticles, and heterogeneous catalysts. A microscope with a sub-angstrom probe size and a sub-100 meV energy resolution can have a tremendous impact on the fundamental understanding of heterogeneous catalysis and on developing novel nanostructured catalysts.

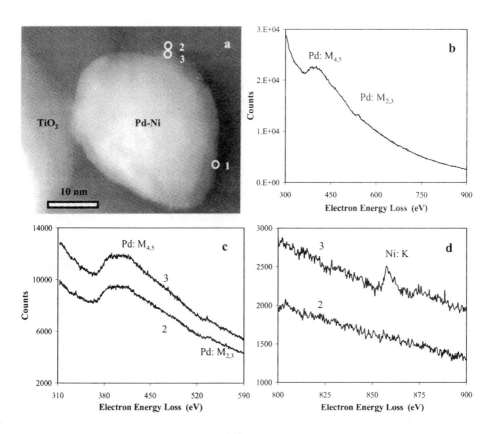

FIGURE 18.12. HAADF image (a) and EELS spectra of a Pd-Ni/TiO$_2$ bimetallic catalysts show alloying of Pd with Ni and the preferential surface segregation of Pd (Reprinted with the permission of the Cambridge University Press and the Microscopy Society of America).[20]

18.4.5. Electron Nanodiffraction Study of Supported Catalysts

When an electron nanoprobe is stopped at any point of interest, for example, a Pd nanoparticle shown in Fig. 18.1b, a convergent beam electron nanodiffraction pattern is formed on the detection plane of a STEM instrument. The size of the diffraction spots is determined by the convergence angle of the incident electron probe; a small electron probe generally requires a large convergence angle. With a two-dimensional detector such as a phosphor screen or a CCD (charge-coupled device) system, the nanodiffraction patterns can be observed, recorded, and analyzed in the same away as convergent beam electron diffraction (CBED) patterns obtained in TEM instruments.[59-62]

The use of a field-emission gun in STEM, however, introduces important new features in STEM CBED patterns.[59] First, the sizes of the electron probes are usually 1 nm or less in diameter, much smaller than those used in TEM. Second, the use of a field-emission gun warrants the coherent nature of a convergent nanoprobe: the illuminating aperture is filled with completely coherent radiation and the final probe entering the specimen can be treated as perfectly coherent. In contrast, the illuminating aperture in conventional TEM is considered incoherently filled and the illumination is treated as completely incoherent. Coherent electron nanodiffraction (CEND) is the only technique that gives full diffraction information on individual nanoparticles. Figure 18.13a shows a schematic diagram illustrating a coherent, convergent electron nanoprobe with a size smaller than the size of a multi-faceted nanoparticle. Diffraction patterns from the various parts of the nanoparticle can be obtained to give information about the structure as well as the morphology of the nanoparticle.

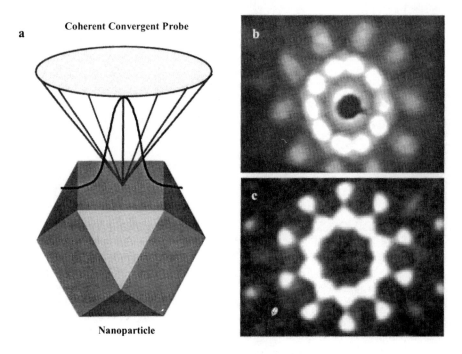

FIGURE 18.13. Schematic diagram (a) illustrates the setup of obtaining a coherent electron nanodiffraction pattern from a nanoparticle. (b) and (c) show experimental and simulated electron nanodiffraction patterns, respectively, from an icosahedral Ag nanoparticle revealing five-fold symmetry of the crystal structure (Reprinted with the permission of WILEY-VCH).[11]

Although CEND patterns obtained in a STEM instrument are necessarily CBED patterns, different operating modes of STEM require different incident beam convergent angles. The spatial distribution of the electron probe depends on the convergence angle of the incident beam. For example, high-resolution lattice imaging requires the smallest electron probe and overlapping diffraction spots; thus, a large convergence angle of the incident probe is used to satisfy these conditions. On the other hand, it is necessary to use a small convergence angle to obtain sharp diffraction spots from small particles or localized crystal defects. If we denote the convergence semi-angle of the incident probe as α and the Bragg angle of a crystal as θ_B, then three distinctive types of CEND patterns can be formed: 1) $\alpha \leq \theta_B$; 2) $\alpha > \theta_B$; and 3) $\alpha \gg \theta_B$.[11] For characterizing heterogeneous catalysts, especially supported nanoparticles, we usually use smaller convergence angles to obtain non-overlapping and relatively sharp diffraction spots to provide structural information.

When metal atoms aggregate from the vapor phase or in a liquid, they usually form a crystal, having shapes of regular pentagonal bi-prisms or icosahedra. Their internal structure is a complex arrangement of five or twenty twinned components. Large metal particles with shapes of cub-octahedron, decahedron, icosahedron, and other multiple-twinned structures have been observed.[34] For particles with sizes smaller than 2 nm in diameter, however, it is difficult to unambiguously determine their shapes by HREM imaging techniques.

CEND technique can provide information about the shape of clean, metallic nanoparticles provided these nanoparticles are stable under the electron beam irradiation. For example, a large fraction of clean silver nanoparticles with sizes smaller than 3 nm in diameter was observed to give unique CEND patterns exhibiting five-fold-symmetry. These clean silver nanoparticles were formed by in situ deposition in a UHV STEM instrument. Figure 18.13b shows such a CEND pattern and Fig. 18.13c shows a simulated CEND pattern of a small Ag icosahedron with the incident beam direction along the five-fold-symmetry axis. The simulated CEND pattern closely matches the experimental one. Detailed analyses of CEND patterns recorded on videotapes can provide information about the shape of, as well as the defect structure in, nanoparticles. The wide application of this powerful diffraction technique is currently limited by the availability of the special attachments that make it easy to record nanodiffraction patterns.

One of the successful applications of STEM techniques is the ability to analyze the composition as well as the structure of individual nanoparticles. Figure 18.14, for instance, shows a set of nanodiffraction patterns, obtained with an electron beam of approximately 1 nm in diameter, of a PtPd bimetallic catalyst containing Pt, Pd, Pt-Pd nanoparticles supported on zeolite crystals which are connected by alumina binders. These CEND patterns provide information about the crystallographic structure of the individual bimetallic nanoparticles and supports, and information about the structural relationship between the metal particles and their supports or binder materials.

It is, however, impossible to make accurate measurements of lattice parameters in CEND patterns because of the large sizes of the diffraction spots. An error of 5% or higher is common in determining lattice constants of small particles, and much larger errors can frequently occur because of the coherent interference effects.

It is important to correlate the characteristic features of CEND patterns to particle properties, such as the structure of the particle, the nature of defects within the particle, or the shape and size of the particle. A frequently observed characteristic feature is the splitting of diffraction spots along certain crystallographic directions (e.g., see diffraction pattern number 4 in Fig. 18.14). The spot splitting in non-overlapping CEND patterns is attributable to the coherent nature of electrons diffracting from an abrupt discontinuity of the scattering potential at particle edges.[63] It is also observed that the spot splitting is related to the geometric forms of the diffracting particles; some splitting occurs in a well-defined crystallographic direction. Depending on the probe position relative to the center of a particle, annular rings may be observed in CEND patterns of small particles. Dynamical simulations reveal that for a particle that has facets smaller than the size of the incident probe, the incident electrons may interact with several facets of the small particle.[64] The thickness of the particle may vary rapidly even within a region of only about 1 nm in diameter. The electron probe effectively interacts with the "particle morphology" under illumination. The intensity variations of the splitting spots are related to the probe positions with respect to the particle facets and are related to the length of the facets along the incident beam direction.

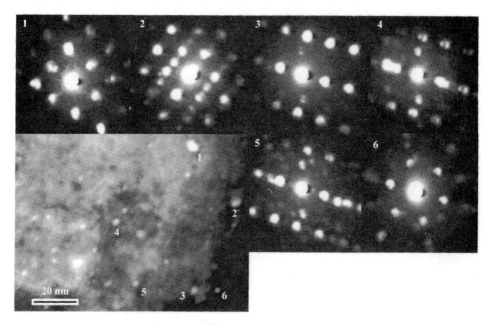

FIGURE 18.14. HAADF image of a Pd-Pt/zeolite catalyst and the corresponding nanodiffraction patterns obtained from the individual nanoparticles (Reprinted with the permission of WILEY-VCH).[11]

18.4.6. Auger Electron Spectroscopy and Scanning Auger Microscopy of Surfaces

In a STEM instrument, Auger electrons, emitted from either the entrance or the exit surface of a specimen, can be collected and analyzed using a CMA (cylindrical mirror analyzer) or a CHA (concentric hemispherical analyzer) electron spectrometer. Because of the high energy and high brightness of the incident electrons, the employment of magnetic "parallizers", and the use of thin specimens in STEM instruments, high-quality Auger electron spectra can be acquired with high peak-to-background ratios.[65-67] High-energy resolution AES spectra of clean silver nanoparticles with sizes in the range of 1-5 nm in diameter have been obtained.[65, 66]

Surface compositional analysis of individual nanoparticles is critical to understanding the performance of industrial bimetallic or multi-component catalysts used in a variety of chemical reactions. The composition of individual nanoparticles can usually be obtained by the use of high spatial resolution XEDS. It is, however, extremely difficult to extract information on the preferential surface segregation or aggregation of individual components in nanoparticles of different sizes. Because of the high-surface sensitivity of Auger electrons, it is possible to determine qualitatively and, in some cases, quantitatively, the surface composition of nanoparticles consisting of multiple components. Figure 18.15 shows an Auger electron spectrum obtained from a sample containing palladium, silver, or palladium-silver nanoparticles highly dispersed on a high-surface area alumina support. Both the silver and the palladium MNN Auger peaks are clearly revealed. Quantitative analyses of this type of spectra provided information on the surface enrichment of specific elements as well as information about how this preferential surface segregation varies with the size of the individual nanoparticles in supported catalysts.

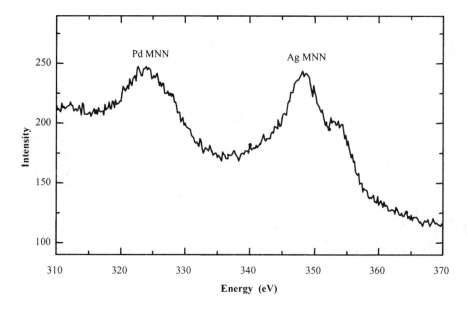

FIGURE 18.15. Auger electron spectrum obtained from a Pd-Ag/alumina model catalyst clearly shows the Ag and Pd MNN Auger peaks with good signal-to-noise and signal-to-background ratio (Reprinted with the permission of WILEY-VCH).[11]

With the improvement in the detection efficiency of low-energy electrons in the dedicated STEM system by employing magnetic "parallelizers",[68] an image resolution < 1 nm can be achieved in scanning Auger microscopy (SAM) images of small nanoparticles; for example, silver nanoparticles < 1 nm in diameter and containing as few as 15 silver atoms can be detected.[65]

The digitally acquired Auger peak and background signal images contain quantitative data on the Auger peak height and SE background intensity. Data sets obtained by processing the digital images contain quantitative information about the contrast of the Auger peak and background signal images.[66]

Although it is difficult to image the surface composition of nanoparticles in heterogeneous catalysts, nanometer resolution SAM images of a bimetallic catalyst has been obtained. Figure 18.16a shows a HAADF image of a zeolite-supported PdM (M: another transition metal) bimetallic catalyst, revealing the presence, size and spatial distribution of the small metal or alloy nanoparticles. The HAADF image, however, cannot provide any information about the surface composition of the individual nanoparticles. On the other hand, the Pd Auger elemental map of Fig. 18.16b, however, clearly shows that some of the small nanoparticles are located on the surface of the zeolite support and that the surface composition of these alloy nanoparticles is rich in Pd. These nanometer resolution elemental maps provide useful information on the surface chemical composition of heterogeneous catalysts and on how the surface composition of the individual nanoparticles changes after catalytic reactions.

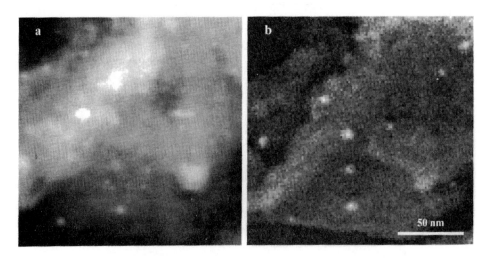

FIGURE 18.16. HAADF image (a) and Pd MNN Auger map (b) of a PdM/zeolite (M: second metal) catalyst.

The resolution in SAM images depends on several sample- and instrument-related effects. The sample-related effects include: 1) surface topography, 2) escape depth of the collected Auger electrons, 3) contribution from backscattered electrons, and 4) localization of the Auger electron production processes. The last factor sets the ultimate resolution limit that is achievable in SAM images. Since the primary inelastic scattering processes involve excitation of inner-shell electrons, the generation of Auger electrons is highly localized. With thin specimens and high-energy incident electrons, the contribution from backscattered electrons should be negligible or significantly reduced; it may degrade the image resolution and affect the image contrast of bulk samples.

The instrument-related effects include: 1) the intensity distribution of high-energy electron probes, 2) the collection efficiency of the emitted Auger electrons, and 3) the instability of the STEM microscopes. At present, the instrument-related factors set the limits of obtainable resolution to about 1 nm in Auger peak images of thin specimens.

Because the intensity profile of a particle in a SAM image is a complicated convolution of the electron probe with the real morphology of the particle, the true size of the particle may be smaller than that measured in a SAM image. A de-convolution process might be used to extract the real sizes of the observed nanoparticles. However, knowledge of the intensity distribution of the probe, the radial and angular distributions of the emitted Auger electrons, and the detailed transmission function of the electron analyzer would have to be known.

The emission of Auger electrons from small particles is different from that of flat surfaces. Electrons generated inside a small particle may all escape if the radius of the particle is much smaller than the inelastic mean free path (λ) of the collected Auger electrons. In this limit, the total number of the collected Auger electrons is proportional to the volume of the particle; and the distinction between "bulk" and surface signals is no longer valid. A quantitative estimate of the minimum particle size detectable in SAM images can be developed from a simple relationship between the signal strength and the size of the particle. By using an intensity-ratio method, it is possible to estimate the total number of atoms detectable in high spatial resolution Auger maps.[65] On this basis, it is estimated that particles containing as few as 15 silver atoms can be detected in high resolution SAM images. At present, the minimum detectable mass in high spatial resolution SAM images is less than 3×10^{-21} g.[65]

The minimum detectable dimension is different from the resolution of SAM images. While the latter is currently limited by the sizes of the incident electron probe, the former is related to, but not limited by, the dimension of the incident electron probe. The minimum detectable size can be much smaller than the incident probe diameter. The minimum particle size detectable in a SAM image is directly related to the signal-to-noise and signal-to-background ratios, as well as the radiation damage of the sample. For example, the minimum detectable dimension would be the size of a single atom if the signal strength were not a limiting factor and that the atom were stable under intense electron beam irradiation.

18.5. HIGH-RESOLUTION SCANNING ELECTRON MICROSCOPY OF SUPPORTED CATALYSTS

Although TEM/STEM based techniques can provide important information about heterogeneous catalysts with an atomic-scale resolution, a major limitation of these techniques is, however, the stringent requirement of samples that can be examined: useful information can be extracted only from very thin areas of a sample because of severe absorption of electrons in thicker regions. The preparation of suitable samples may also pose a major problem for observing certain types of catalysts, for example, catalyst beads, pellets, powders, or cylinders that are frequently used in industrial catalytic reactions. These limitations preclude the application of TEM/STEM techniques to extracting information about the surface properties of thick or bulk catalyst samples.

Fortunately, the recent rapid development of the high-resolution scanning electron microscopy (SEM) makes it possible to examine bulk samples on a nanometer or sub-nanometer scale.[69] The high-stopping power of electrons at low energies significantly reduces the electron-specimen interaction volume as well as the range that an electron can penetrate into the sample. Low-voltage SEM (LVSEM) offers new opportunities for imaging a wide variety of bulk samples with high spatial resolution.[70] Highly dispersed metal nanoparticles as well as detailed surface topography of supported catalysts have been examined.[41, 71-74]

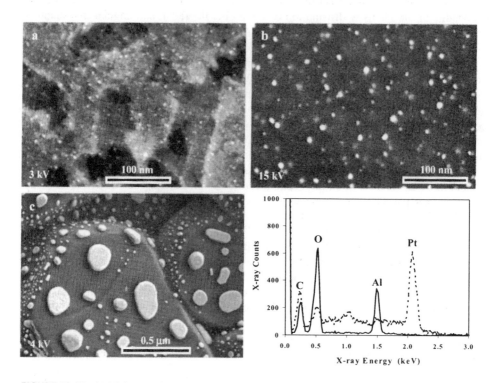

FIGURE 18.17. (a) High-resolution SE image of a Pt/carbon catalyst shows Pt nanoparticles and the carbon pore structure. (b) High-resolution backscattered electron image of a Pt/carbon catalyst shows the Pt nanoparticles with high atomic number contrast. (c) SE image of a Pt/alumina model catalyst shows the different sintering behavior of Pt nanoparticles on different faces of the alumina support. (d) Low-voltage XEDS spectra obtained from the Pt particles (dotted line) and the alumina support (solid line) shown in the SE image c) (Reprinted with the permission of the Cambridge University Press and the Microscopy Society of America).[20]

Figure 18.17a is a high-resolution SE image of a carbon-supported Pt catalyst, clearly showing the detailed pore structure and the general morphology of the carbon support. The small Pt nanoparticles are also clearly revealed. Detailed contrast mechanisms of these small metal nanoparticles have been fully investigated.[75] The application of a bias voltage to the sample can further tune the contrast of the small metal nanoparticles.[76] Because of the complicated electron beam interactions with the "embedded" metal nanoparticles located near the sample surface, quantitative interpretation of the particle contrast of high-resolution LVSEM images, such as the ones shown in Fig. 18.17a, is not straightforward.[75] High-resolution SE images, nevertheless, can provide information on the size and spatial distribution of the metal nanoparticles as well as their spatial relationship with respect to the pores of the support material. For example, the brighter dots shown in Fig. 18.17a represent those Pt nanoparticles that are located on the external surface of the carbon powder; the dimmer dots represent the Pt nanoparticles that are located within the carbon support.

With the use of a high-sensitivity backscattered electron (BE) detector, high resolution BE images of metal nanoparticles can be observed with high atomic-number contrast. Figure 18.17b shows such a BE image obtained with an incident beam voltage of 15 KV. The contrast of the metal nanoparticles in BE images has been explained based on Monte Carlo simulations.[75] Again, the brighter dots represent the Pt nanoparticles that are located on or very close to the carbon surface. Generally speaking, the dimmer dots represent those metal nanoparticles that are located inside the support. The particle contrast strongly depends on the distance from the center of the particle to the entrance surface of the electron beam.[75] Images of the same region obtained with various electron beam energies can provide the three-dimensional distribution of the metal nanoparticles. With the development of the newer generation FEG SEM instruments, sub-nanometer resolution can be achieved in both SE and BE images of supported catalysts.

One of the main advantages of using a modern FEG-SEM is its versatility: different imaging modes can be used simultaneously and various beam voltages can be used without significantly altering the image resolution. Thus, low-voltage, high-resolution images can be obtained even for non-conducting materials. For example, Fig. 18.17c shows a SE image of an alumina-supported Pt model catalyst. The model catalyst was prepared by depositing Pt nanoclusters onto pretreated polycrystalline alumina supports; the model catalyst was then reduced at various temperatures in order to study the sintering behavior of supported Pt nanoparticles. Figure 18.17c clearly shows that the sintering of the Pt nanoparticles strongly depends the crystallographic face and the surface morphology of the polycrystalline alumina support. The Pt sintering process was clearly affected by its mobility on the different faces of the alumina support. On certain faces of the alumina powders, the Pt nanoparticles moved easily during the reduction process to form large, flat islands. On other surfaces, however, the Pt nanoparticles did not sinter much at all. The Pt nanoparticles interacted strongly with surface steps and other defects on the alumina support. This type of information is critical to designing catalysts with desired catalytic properties and long-term stability.

High-resolution chemical microanalysis of bulk samples can also be realized at low voltages by collecting and analyzing the emitted characteristic X-ray signals[77, 78] The XEDS spectra shown in Fig. 18.17 demonstrates the use of this microanalysis technique to identify features of interest revealed in either the SE or the BE images. The spectrum represented by the dotted line was obtained from one of the larger Pt islands. The solid line represents a XEDS spectrum obtained from the nearby alumina support. These spectra clearly demonstrate the applicability of high spatial resolution microanalysis with low-energy electrons. Monte Carlo simulations suggest that a lateral resolution of about 50 nm or less should be achievable with 4-keV electrons. The combination of high-resolution SE and BE imaging with low-energy X-ray microanalysis makes the FE-SEM a truly powerful tool for providing high spatial resolution morphological, structural, and chemical information on a wide variety of heterogeneous catalysts that cannot be easily examined in TEM or STEM instruments.

18.6. RECENT AND FUTURE DEVELOPMENTS

Although the electron microscope was invented over 70 years ago the field of high-resolution electron microscopy and spectroscopy is still rapidly progressing. With the development of better and more stable field-emission-guns, the various high-sensitivity detectors, and most importantly the ever-increasing power of desktop computers, a modern electron microscope is becoming more automated and user-friendly; and even remote-microscopy is becoming a reality. On another frontier, the electron microscopy community has been continuously pursuing the goal of sub-angstrom spatial resolution imaging, sub-100 meV energy resolution spectroscopy, the development of novel imaging and analytical techniques, and the development of specialized specimen chambers or stages for in situ studies. In the following, we briefly discuss the very recent development of advanced electron microscopy techniques and the application of these techniques to the characterization of nanoparticles and heterogeneous catalysts.

18.6.1. Atomic Resolution Environmental Electron Microscopy

The art of catalysis has for many years been difficult to be correlated to the surface science studies and this fact is primarily related to the gaps between surface science and catalysis: the materials gap; the pressure gap, and the complexity gap. Catalytic reactions usually occur under high gas pressure or in a liquid; and whatever the initial structure of a catalyst, it will change upon the starting of the catalytic reaction.

A catalyst is not a true catalyst until it has been subjected to the catalytic reaction conditions. Catalysis practitioners all know that the performance of a heterogeneous catalyst at the beginning of the catalytic reaction does not reflect the "equilibrium" state of the catalyst. Therefore, it has been a major concern in the catalysis community about the scientific relevance of results obtained under unrealistic conditions such as single crystals under ultra-high-vacuum (UHV) conditions. To bridge this so-called pressure gap, many in situ spectroscopy techniques have been developed and used to provide information on the dynamic behavior of catalytic reactions and the evolution of heterogeneous catalysts during the chemical reactions.[79] Although these spectroscopic techniques have significantly enhanced our knowledge about heterogeneous catalysis they cannot provide atomic scale information on the behavior of individual active sites. In situ scanning probe techniques, especially STM techniques, can provide atomic scale images; but these techniques cannot provide any compositional information and cannot be applied to practical industrial catalysts. Although ex situ HREM techniques have contributed significantly to the fundamental knowledge of heterogeneous catalysts, the catalyst samples are usually analyzed under static conditions and under high vacuum that are often not representative of the true dynamic state of a working catalyst.

To better understand the dynamic active sites, to gain knowledge on the complex structural changes, and to elucidate catalytic reaction mechanisms, it is desirable to directly observe the structural and chemical evolution of heterogeneous catalysts under dynamic conditions. The recent advancement in developing atomic resolution environmental electron microscopy has made it possible to directly examine nanostructural evolution of heterogeneous catalysts under reaction conditions and has significantly expanded our knowledge on the atomic structure of heterogeneous catalysts under realistic conditions.[80-83] Catalyst samples can now be examined at temperatures between −196 °C and > +1000 °C at elevated gas pressures of a variety of gases. Furthermore, even liquid catalytic reactions can be examined with the use of special sample holders. For example, the nanoscale imaging and electron diffraction of dynamic liquid hydrogenation and polymerization reactions in the manufacturing of polyamides were recently reported.[83] High-resolution studies of a wide range of solution-solid and solution-gas-solid reactions can now be achieved in environmental HREM (E-HREM) instruments.

E-HREM technique has already provided important information on a variety of heterogeneous catalyst systems including strong-metal support interactions (SMSI) in the Pt/TiO$_2$ system,[80] selective oxidation of n-butane to maleic anhydride over vanadium phosphorus oxides,[80] hydrogenation reactions in nylon 6,6 over novel xerogel catalysts,[83] and polymerization reactions over Ziegler-Natta catalysts.[84, 85]

E-HREM technique has also been applied to the fundamental study of the action of a promoter on a heterogeneous catalyst.[86]   In this study the authors demonstrate that atomic scale structural characterization of a catalyst's surface in the presence of reaction gases can help clarify how a catalyst modifier promotes the catalyst's activity. The increased activity of a barium-promoted boronnitride-supported Ru catalyst is related to a two-dimensional barium-oxygen overlayer on the ruthenium nanocrystals. The authors also show that conventional ex situ TEM fails in this case because the surface structure of the catalyst is completely different in vacuum. In another recent study the same research group investigated, using E-HREM technique, the dynamic shape changes of a zinc oxide-supported copper catalyst system, which is used industrially for methanol synthesis.[87] The authors found that the copper nanocrystals undergo dynamically reversible shape changes in response to changes in the gas environment. By analyzing the detailed atomic scale structural changes, the authors proposed new insights into the adsorbate-induced structural changes of metal nanocrystals. The observed dynamic restructuring of the catalyst also directly demonstrates that the relevant active sites are generated during the catalytic reactions.

The wide application of E-HREM to studying heterogeneous catalysts will undoubtedly bridge the pressure gap between the surface science and the real-world heterogeneous catalysis. The information provided by atomic resolution E-HREM can help us understand the structural evolution of the individual nanocomponents during the catalytic reactions or during the catalyst preparation processes. The E-HREM studies can provide important information on the concentration of different types of surface sites such as steps, kinks, corner atoms, and other defects under various gas environments. When combined with the atomic resolution analytical techniques we discussed in this chapter, the E-HREM can also provide information on adsorbate-induced surface segregations in bimetallic catalysts and how these preferential surface segregations vary with the change of gas environment. Advanced E-HREM techniques will become more and more important in understanding the fundamental mechanisms of nanostructured heterogeneous catalysts.

18.6.2. Sub-Angstrom Resolution with Cs-correction

Atomic resolution imaging and spectroscopy in the electron microscope require an electron probe of the size of one atom; the electron probe should also contain enough current for rapid acquisition of electron energy-loss spectra with meaningful signal-to-noise ratio. Because of the unavoidable aberrations of round electron lenses, the practical probe sizes achievable are usually much higher than the electron wavelength. Although aberration correction was proposed more than 60 years ago to overcome the barrier of lens defects[88, 89] it is only within the last few years that practical resolution improvement has been achieved in the SEM,[90] in the TEM,[91, 92] and in the STEM.[93, 94] The recent success of incorporating Cs-correctors into the modern FEG TEM or STEM instruments is, at least in part, attributable to the availability of faster computers which can provide on-line automatic aberration diagnosis and auto-tuning.

The attainment of sub-angstrom probe size with high total beam current in the dedicated STEM instruments opened a new world for applying atomic resolution analytical electron microscopy to the study of heterogeneous catalysts. For example, the movement and diffusion processes of single atoms can be easily examined in practical samples and the shape of small nanoclusters can be determined and analyze.[45] Most importantly, the local electronic structure of steps, kinks, and different facets of individual nanoparticles may be analyzed in the Cs-corrected STEM instruments.

The fundamental limits on the performance of Cs-corrected electron microscopes are now determined by the chromatic aberration and higher order geometric aberrations. When these limitations are overcome, high-brightness electron probes with a probe size < 0.05 nm should be achievable. With sub-angstrom resolution electron microscopes, the routine performance of such instruments is then critically dependent on the control of environmental influences. To routinely achieve ultrahigh resolution, instabilities caused by fluctuations in electromagnetic fields, electrical interference, mechanical vibration, sound, thermal variations, airflow in the electron microscope room, etc. have to be eliminated or significantly reduced. Therefore, future advanced electron microscopes will require special housing facilities and most probably will be remotely controlled.

18.6.3. High-Energy-Resolution Electron Energy-Loss Spectroscopy with Monochromators

Although near-edge fine structures in an EELS spectrum contain information on the chemical bonding of the probed area, the relatively poor energy resolution achievable in the current electron microscopes prevents important applications of this powerful technique to the study of heterogeneous catalysts and other nanophase materials. The energy resolution in the EELS spectrum is determined by the initial energy distribution of the electron source in the electron microscope. For cold field emission sources, resolutions of 0.3 eV are possible; thermally assisted field emitters (e.g., Schottky emitters) give 0.6-0.8 eV; heated LaB6 sources give 1.5 eV; and tungsten filament sources have a resolution of about 3 eV. On the other hand, in XANES spectra an energy resolution < 0.1 eV can be easily achieved.

To improve the energy resolution in the EELS spectra and to reduce the effect of chromatic aberration on the spatial resolution, the electron microscopy community is rapidly developing electron monochromators to be used in the modern FEG TEM or STEM instruments.[95, 96] The incorporation of an electron source monochromator in an electron microscope to reduce the energy width of the primary beam to below 0.1 eV will not only increase the energy resolution of EELS but also enhance the performance of HREM and low-voltage SEM. The combined use of Cs-correctors and monochromators in a STEM instruments should be ideal for studying the electronic structures of heterogeneous catalysts with atomic scale resolution.

The realization of the next generation electron microscopes with sub-angstrom spatial resolution and sub-0.1 eV energy resolution will undoubtedly have significant impact on our fundamental understanding of heterogeneous catalysis. Chemical bonding and atomic coordination information may be attainable on individual nanoparticles. When these new developments are incorporated into the environmental electron microscopes, we may be able to directly probe the adsorption behavior of single molecules and to study industrial catalysis on a molecular level.

18.6.4. Electron Tomography

Electron microscopy, regardless the image resolution and imaging mode used, generates two-dimensional information by projecting through a three-dimensional structure. This fundamental limitation of electron microscopy techniques sometimes prevents us extracting useful information about the samples, especially heterogeneous catalysts and nanophase materials. The highly complex morphologies of industrial heterogeneous catalysts provide most efficient and stable catalysts. Rough surfaces, tortuous and interconnected pores, unique spatial distributions of nanoparticles within the support are all strategies for increasing the surface area, enhancing the selectivity of the catalytic reaction, or prolonging the catalyst stability. The densely packed nature of the nanocomponents in heterogeneous catalysts often leads to the overlap of many features when viewed in an electron microscope; it is often very difficult and most of the time impossible to make conclusions about the three-dimensional structure of the catalysts. To develop nanostructured heterogeneous catalysts with high efficiency and selectivity, we need to design, synthesis, and characterize their three-dimensional architecture based on the nanoscale building blocks. Electron tomography in TEM or STEM allows us to perform three-dimensional analysis of nanostructures in heterogeneous catalysts.

Electron tomography is similar to that of an X-ray CT (computed tomography) scanner in which a tilt series of micrographs is acquired and used to reconstruct the three-dimensional structure of an object. Electron tomography practiced in TEM is a standard tool for the analysis of macromolecules and complex biological ultrastructures with nanometer resolution.[97] Although the bright-field tomography technique can be successfully applied to biological samples that are often amorphous materials, this technique, however, has limited applications to crystalline materials because the intensity in a micrograph strongly depends the orientation of a crystal.

The recent development of electron tomography using HAADF imaging has made it possible to directly applying this technique to studying the three-dimensional structures of nanostructured catalysts such as Pd/C and Pd_6Ru_6/MCM-41 supported catalysts.[98-100] Alternatively, inelastically scattered electrons can also be used as input signals to form a series of element specific images that are used to reconstruct three-dimensional objects.[101]

Electron tomography has been used in the biosciences for over 30 years and has proved invaluable for understanding the complex macromolecules and other biological structures. Successful applications of electron tomography to materials science, especially to heterogeneous catalysts, are, however, realized only in the last 2-3 years. Although much more development work has to be done before routine reconstructions of catalysts can be realized, it has already been demonstrated that high-resolution electron tomography can provide important information about the complex three-dimensional catalyst structures such as the true spatial distribution of metallic nanoparticles and agglomeration of nanoparticles.

To routinely apply electron tomography technique to characterizing heterogeneous catalysts the following two issues need to be solved: 1) automatic acquisition of datasets with high spatial resolution and precision, 2) use of special stages that allow large tilting angles. With the use of the Cs-correctors, large gaps between the objective pole pieces can be used without sacrificing the image resolution; therefore, the second problem can be addressed in the newer generation of electron microscopes. With the rapid progress in the automation of modern electron microscopes and the development of new algorithms and 3D reconstruction image analysis softwares the first problem should also be overcome in the near future. When Cs-corrected electron microscopes are used and automatic high-resolution electron tomography technique is developed we expect to extract information not only on the 3D distribution of nanoparticles in supported catalysts but also the shapes of the individual nanoparticles and their relationship with respect to the support materials.

18.6.5. Electron Holography

In conventional HREM, the phase and amplitude images of the object are mixed and convoluted with the phase contrast transfer function of the objective lens; therefore, the retrieval of the phase and amplitude images are difficult to achieve. Electron holography, proposed by Dennis Gabor as early as 1948,[102] is a technique that can provide information on both the phase and the amplitude of the imaged object. Holography is based on the principle of interference of electron waves without the distortion introduced by the use of imperfect lenses. The development of high-brightness and high-coherent field-emission electron guns made it possible to obtain electron holograms in the modern FEG TEM/STEM instruments. There are many ways to perform electron holography in the modern TEM or STEM instruments.[103]

Electron holography technique has been extensively used to image electrostatic field and charge distributions in transistors and ferroelectric materials, to study magnetic domains and flux lines in superconducting materials and magnetic materials, and to improve image resolution.[104-107] Electron holography is still the only method for investigating magnetic fields and electric potentials on a nanometer scale. Although posteriori correction of coherent wave aberrations in electron holograms can provide better image resolution and help interpretation of high-resolution images, direct sub-angstrom resolution imaging by using aberration-corrected electron microscopes may prove to be more accessible and user-friendly.

The application of electron holography technique to characterizing heterogeneous catalysts is limited, primarily due to the complexity of industrial heterogeneous catalysts. This technique, however, has been used to determine the morphology of nanoparticles because the pure phase information in the image wave can be precisely reconstructed and displayed.[108, 109] Intensity profiles over the phase image are directly related to the thickness and in some special cases are directly related to the particle morphology. Electron holography technique also revealed the presence of internal voids within single crystal Pd nanoparticles in a supported Pd catalyst.[110] The ability to determine particle shapes and surface facets should be of great interest as a method for characterizing heterogeneous catalysts.

18.7. CHALLENGES AND OPPORTUNITIES

In the above sections we described the applications of several advanced electron microscopy techniques to characterizing heterogeneous catalysts; each technique can provide useful and complementary information. The added value of applying electron microscopy to understanding the properties of catalysts, however, often comes from employing the combination of all or a few selected techniques.

The successful applications of advanced electron microscopy techniques to catalyst characterization have made these techniques indispensable for systematically developing industrial heterogeneous catalysts. Because of the nature of high spatial resolution techniques, however, we need always keep in mind that electron microscopy results generally have extremely poor statistical value. For example, in a supported metal catalyst used in commercial reactors, there are about 1-100 trillion metal nanoparticles per cm^3; and an electron microscopist can only analyze a minute portion of the catalyst. Whenever possible, the microscopy data should be supplemented by other broad-beam or bulk techniques. The application of advanced electron microscopy techniques to characterizing heterogeneous catalysts usually consists of four stages: 1) proof of concept, 2) qualitative description, 3) quantitative description, and 4) statistically useful data. The last stage is the most valuable to developing industrial catalysts; and it is the most challenging for an electron microscopist. In the following, we discuss some challenging areas and also point out some new research initiatives that may significantly benefit the fundamental understanding of heterogeneous catalysis.

18.7.1. Size Distribution of Nanoparticles

The knowledge on the accurate size distribution of metal nanoparticles is critical to developing supported metal catalysts with desired properties. This aspect has been clearly demonstrated recently with the discovery of nano-gold as an excellent oxidation catalyst at low temperatures.[111] Only if the size of the gold nanoparticles is in the narrow range of 2-4 nm then the titania-supported gold nanocatalyst gives surprisingly high activity; deviations from this size range give no significant activity at all. Thus, the performance of the gold nanocatalyst depends strongly on the control of the size distribution of the gold nanoparticles.

When we develop nanostructured catalysts, the concept of the "average size" of nanoparticles is misleading since numerous size distributions can give the same average size. Instead, we should use size distribution to describe supported metal catalysts. It is the size distributions, not the average sizes, which correlate to the performance of the nanostructured catalysts.

Accurate determination of the size distribution of small metal nanoparticles in industrial catalysts, however, proves to be a very challenging task. We do not have a robust method to provide statistically reliable size distributions of very small nanoparticles. We also need reliable, high-throughput techniques since many catalyst samples have to be analyzed if one wants to develop a successful commercial catalyst. It is also important to note that frequently electron microscopists miss out the imaging and analysis of very large particles. Even though the large particles usually do not contribute much to the performance of the catalyst, these large particles waste a large portion of the active ingredients, thus resulting in a "bad" catalyst. For example, the weight or volume of a 1-μm particle equals that of 1000,000,000 1-nm small nanoparticles altogether. Thus, the statistically meaningful counting of larger particles is extremely important for evaluating the performance of a commercial catalyst.

The most challenging aspect of determining particle size distributions is to correlate variations in size distributions of different samples to their preparation parameters, history of treatment, and catalytic performance. For this type of study, it usually requires measurement of tens of thousands of nanoparticles for each catalyst sample. And up to now we do not have reliable methods to perform this type of analysis automatically and with high throughput. Any breakthrough in reliably and rapidly determining the size distribution of nanoparticles can significantly contribute to the rapid and successful development of industrial heterogeneous catalysts.

18.7.2. Spatial Distribution of Nanoparticles

Only electron microscopy techniques can provide spatial distribution of nanoparticles in industrial heterogeneous catalysts. However, these are no robust protocols that can quantitatively describe the variations in spatial distribution of metal or alloy nanoparticles in supported catalysts. The demand for quantitative description of the spatial distributions of nanoparticles will become stronger and stronger as we move into the era of nanoscience and nanotechnology. The recent application of electron tomography to supported catalysts is clearly a major step forward to describing the three-dimensional distribution of metal nanoparticles in supported catalysts.[99] Future efforts should focus on developing robust and high-throughput techniques to quantitatively determine the spatial distribution of active components in heterogeneous catalysts and to correlate the variations in spatial distributions of the active components to the change in catalytic performances.

18.7.3. Compositional Profile and Surface Segregation of Individual Alloy Nanoparticles

To successfully develop nanostructured bimetallic or multimetallic alloy catalysts, it is indispensable to accurately determining the composition of individual nanoparticles. The composition-size-plot method[48] has proved extremely important for developing industrial bimetallic or multimetallic catalysts. However, in developing industrial catalysts, we usually need to analyze many different samples to correlate the structural variations with the variations in the catalytic performance and catalyst preparation parameters. For each sample we usually need to analyze hundreds to thousands of individual nanoparticles in order to have statistically meaningful results. Thus, any development that can significantly shorten the analysis time or make the analysis processes fully automatic will undoubtedly expand the application of this powerful method to developing industrial heterogeneous catalysts.

Study of surface segregation of individual alloy nanoparticles is still a frontier area of research; its feasibility is demonstrated in this chapter. Future development in this area will greatly increase our knowledge of the behavior of alloy nanoparticles and the fundamental understanding of supported alloy catalysts. The ultimate goal is to be able to determine the surface electronic structure of individual facets, steps, kinks, and surfaces of nanoparticles and to correlate the findings to the bonding, adsorption, and dissociation properties of heterogeneous catalysts.

18.7.4. Interfacial Atomic and Electronic Structure

In the last several years, the catalysis community started to realize that the interfaces and the interfacial region play an important role in determining the performance of supported metal catalysts. This is logical since it has been known for a long time that only certain combinations of a metal with a support can provide significant catalytic activity or selectivity. The so-called "support" is not viewed as an inert "spectator" any more. In terms of nanoscience research, we should now view the active components and their immediate surroundings as an integrated system that behaves coordinately during catalysis.

Therefore, catalysis now extends into the interface science of materials; all the techniques developed for the study of interfaces can be and should be applied to the study of nanostructured catalysts. The bonding or adhesion energies of metal nanoparticles to the support materials (metal oxides, sulfides, and carbides) should be fully investigated. We can simply view supported catalyst systems as a special configuration of nanophase materials often having isolated surfaces and interfaces. Not only the surface (i.e. the interface between a gas or liquid phase and a solid face) structure but also the interface structure of small nanoparticles in heterogeneous catalysts should be fully characterized in order to understand the structure-performance relationship of supported catalysts.

18.8. SUMMARY

In this chapter, we illustrated by using examples the recent applications of advanced electron microscopy techniques to characterizing nanoparticles and heterogeneous catalysts; we discussed the recent development of the next-generation electron microscopy techniques; and we outlined some challenges and opportunities of applying advanced electron microscopy techniques to developing nanostructured catalysts. The combination of HREM, HAADF, XEDS, EELS, and nanodiffraction techniques can now be realized in the modern FEG TEM/STEM microscope and can be successfully applied to the study of heterogeneous catalysts, as demonstrated in this chapter. The combination of these TEM/STEM techniques with the techniques available in the modern FEG-SEM instruments significantly expands the usefulness of electron microscopy for solving critical issues in nanostructured heterogeneous catalysts.

The use of Cs-correctors and monochromators in the next-generation electron microscopes will undoubtedly makes electron microscopy techniques indispensable for understanding the fundamental mechanisms of heterogeneous catalysis as well as for developing industrial heterogeneous catalysts. The recent successful development of atomic resolution environmental EM and the wide application of this technology in the near future will provide us information about the nature of catalysis and the structural evolution, on an atomic-scale, of heterogeneous catalysts during catalytic reactions. High-resolution electron tomography, although in its very early stage of development, has already demonstrated its value for providing true 3D structures of complex catalyst systems. The intrinsic nature of high spatial resolution techniques, however, poses the most significant challenge for the electron microscopy community: how to obtain statistically meaningful data with high throughput. With the rapid development of computer technologies, automation of advanced electron microscopes, and development of new image analysis algorithms the problem of "obtaining statistically meaningful data" should be overcome in the near future.

ACKNOWLEDGEMENTS

The author thanks Drs Peter Crozier, Nigel Browning, Jane Tsen, Kai Sun, and Alan Nicholls for helpful discussions or assistance with the electron microscopes and thanks Dr. Nabin Nag for providing catalyst samples.

REFERENCES

1.   D. J. Smith, The realization of atomic resolution with the electron microscope, *Rep. Prog. Phys.* **60**, 1513-1580 (1997).
2.   J. C. H. Spence, *Experimental High-Resolution Electron Microscopy* (Monographs on the Physics and Chemistry of Materials), Oxford University Press, Oxford (1989).
3.   D. B. Williams and B. C. Carter, *Transmission Electron Microscopy: A Textbook for Materials Science*, Plenum, New York (1996).
4.   R. F. Egerton, *Electron Energy-loss Spectroscopy in the Electron Microscope*, Plenum, New York (1996).
5.   D. E. Newbury, D. C. Joy, P. Echlin, C. E. Fiori and J. I. Goldstein, *Advanced Scanning Electron Microscopy and X-Ray Microanalysis*, Plenum, New York (1986).
6.   J. M. Cowley, High resolution studies of crystals using STEM, *Chem. Script.* **14**, 33-38 (1979).
7.   J. M. Cowley, Microdiffraction, STEM imaging and ELS at crystal surfaces, *Ultramicroscopy* **9**, 231-236 (1982).
8.   J. M. Cowley, Scanning transmission electron microscopy and microdiffraction techniques, *Bull. Mater. Sci.* **6**, 477-490 (1984).
9.   J. C. H. Spence and J. M. Cowley, Lattice imaging in STEM, *Optik* **50**, 129-142 (1978).
10.   J. Liu and J. M. Cowley, High-resolution scanning transmission electron microscopy, *Ultramicroscopy* **52**, 335- 346(1993).
11.   J. Liu, Scanning transmission electron microscopy of nanoparticles, in: *Characterization of Nanophase Materials*, Z. L. Wang, ed., Wiley-VCH, New York, pp 81-132 (2000).
12.   J. I. Goldstein, D. E. Newbury, P. Echlin and D. C. Joy, Scanning *Electron Microscopy and X-Ray Microanalysis: A Text for Biologists, Materials Scientists, and Geologists*, Plenum, New York (1992).
13.   M. Jose-Yacaman and M. Avalos-Borja, Electron microscopy of metallic nano particles using high- and medium-resolution techniques, *Catal. Rev.-Sci. Eng.* **34**, 55-127 (1992).
14.   A. K. Datye, and D. J. Smith, The study of heterogeneous catalysts by high-resolution electron microscopy, *Catal. Rev.-Sci. Eng.* **34**, 129-178 (1992).
15.   P. L. GaiBoyes, Defects in oxide catalysts-fundamental-studies of catalysis in action, *Catal. Rev.-Sci. Eng.* **34**, 1-54 (1992).
16.   G. Ertl, Heterogeneous catalysis on atomic scale, *J. Mol. Catal. A* **3443**,1-12 (2002).
17.   L. D. Marks and D. J. Smith, Direct surface imaging in small metal particles, *Nature* **303**, 316-317 (1983).
18.   K. Sun, J. Liu and N. D. Browning, Correlated atomic resolution microscopy and spectroscopy studies of Sn(Sb)O2 nanophase catalysts, *J. Catal.* **205**, 266-277 (2002).
19.   D. A. Jefferson and P. J. F. Harris, Direct imaging of an adsorbed layer by high-resolution electron microscopy, *Nature* **332**, 617-620 (1988).

20. J. Liu, Advanced electron microscopy characterization of nanostructured heterogeneous catalysts, *Microsc. Microanal.* (in press).

21. K. Heinemann and F. Soria, On the detection and size classification of nanometer-size metal particles on amorphous substrates, *Ultramicroscopy* **20**, 1-14 (1986).

22. J. C. Barry, L. A. Bursill and A. V. Sanders, Electron microscope images of icosahedral and cuboctahedral (fcc. packing) clusters of atoms, *Aust. J. Phys.* **38**, 437-448 (1985).

23. P. L. Gai, M. J. Goringe and J. C. Barry, HREM image contrast from supported small metal particles, *J. Microsc.* **142**, 9-24 (1986).

24. M. H. Yao and D. J. Smith, HREM image simulations for small-particle catalysts on crystalline supports, *J. Microsc.* **175**, 252-265 (1994).

25. S. Bernal, J. J. Calvino, M. A. Cauqui, J. M. Gatica, C. Larese, C. Lopez-Cartes, J. A. Perez-Omil and J. M. Pintado, Some recent results on metal/support interaction effects in NM/CeO₂ (NM: noble metal) catalysts, *Catal. Today* **50**, 175-206 (1999).

26. A. D. Logan, E. Braunschweig, A. K. Datye and D. J. Smith, Direct observation of the surfaces of small metal crystallites: rhodium supported on titania, *Lanngmuir* **4**, 827-830 (1988).

27. A. D. Logan, E. Braunschweig, A. K. Datye and D. J. Smith, The oxidation of small rhodium metal particles, *Ultramicroscopy* **31**, 132-137 (1989).

28. S. Bernal, J. J. Calvino, M. A. Cauqui, G. A. Cifredo, A. Jobacho and J. M. Rodriguez-Izquierdo, Metal-support interaction phenomena in rhodium/ceria and rhodium/titania catalysts: comparative study by high-resolution transmission electron spectroscopy, *Appl. Catal. A* **99**, 1-8 (1993).

29. S. Bernal, F. J. Botana, J. J. Calvino, C. Lopez-Cartes, J. A. Perez-Omil and J. M. Rodriguez-Izquierdo, The interpretation of HREM images of supported metal catalysts using image simulation: profile view images, *Ultramicroscopy* **72**, 135-164 (1998).

30. P. A. Crozier, S.-C. Y. Tsen, C. Lopes-Cartes, J. Liu and J. J. Calvino, Factors affecting the accuracy of lattice spacing determination by HREM in nanometre-sized Pt particles, *J Electron Microsc.* **48**, 1015-1024 (1999).

31. S.-C. Y. Tsen, P. A. Crozier and J. Liu, Lattice measurement and alloy compositions in metal and bimetallic nanoparticles, *Ultramicroscopy* (in press) (2003).

32. S. Ino, Epitaxial growth of metals on rock salt faces cleaved in vacuum II: Orientation and structure of gold particles formed in ultra-high vacuum, *J. Phys. Soc. Japan* **21**, 346-362 (1966).

33. J. G. Allpress and J. V. Sanders, Structure and orientation of crystals in deposits of metals on mica, *Surf. Sci.* **7**, 478-483 (1967).

34. L. D. Marks, Experimental studies of small particle structures, *Rep. Prog. Phys.* **57**, 603-649 (1994).

35. S. Tehuacanero, R. Herrera, M. Avalos and M. Jose-Yacaman, High resolution TEM studies of gold and palladium nano-particles, *Acta. Metall. Mater.* **40**, 1663-1674 (1992).

36. M. Jose-Yacaman, M. Marin-Almazo and J. A. Ascencio, High resolution TEM studies on palladium nanoparticles, *J. Mol. Catal. A* **173**, 61-74 (2001).

37. J. O. Malm and M. A. Okeefe, Deceptive "lattice spacings" in high-resolution micrographs of metal nanoparticles, *Ultramicroscopy* **68**, 13-23 (1997).

38. M. M. J. Treacy, A. Howie and C. J. Wilson, Z contrast of platinum and palladium catalysts, *Phil. Mag. A* **38**, 569-585 (1978).

39. S. J. Pennycook, A. Howie, M. D. Shannon and R. Whyman, Characterization of supported catalysts by high-resolution STEM, *J. Mol. Catal.* **20**, 345-355 (1983).

40. M. M. J. Treacy and S. B. Rice, Catalyst particle sizes from Rutherford scattered intensities, *J. Microsc.* **156**, 211-234 (1989).

41. J. Liu and J. M. Cowley, High-angle ADF and high-resolution SE imaging of supported catalyst clusters, *Ultramicroscopy* **34**, 119-128 (1990).

42. P. D. Nellist, S. J. Pennycook, Direct imaging of the atomic configuration of ultradispersed catalysts, *Science* **274**, 413-415 (1996).

43. A. V. Crew, J. Wall, J. Langmore, Visibility of a single atom, *Science* **168**, 1338-1340 (1970).

44. J. Liu, HAADF imaging of metal nanoclusters and nanoparticles: challenges and opportunities, in *Proceedings of the 15th International Congress on Electron Microscopy*. pp499-500 (2002).

45. P. E. Batson, N. Dellby and O. L. Krivanek, Sub-angstrom resolution using aberration corrected electron optics, *Nature* **418**, 617-620 (2002).

46. K. Sun, J. Liu, N. K. Nag, and N. D. Browning, Atomic scale characterization of supported Pd-Cu/γ-Al₂O₃ bimetallic catalysts, *J. Phys. Chem.* **B106**, 12239-12246 (2002).

47. C. E. Lyman, R. E. Lakis and H. G. Jr. Stenger, X-ray emission spectrometry of phase separation in Pt-Rh nanoparticles for nitric oxide reduction, *Ultramicroscopy* **58**, 25-34 (1995).

48.   C. E. Lyman, R. E. Lakis, H G. Jr. Stenger, B. Totdal, and R. Prestvik, Analysis of alloy nanoparticles, *Mikrochimica Acta* **132**:301-308 (2000).

49.   L. Bednarova, C. E. Lyman, E. Rytter and A. Holmen, Effect of support on the size and composition of highly dispersed Pt-Sn particles, *J. Catal.* **211**, 335-346 (2002).

50.   C. E. Lyman, J. I. Goldstein, D. B. Williams, D. W. Ackland, S. Von Harrach, A. W Nicholls and P. J. Statham, High-performance X-ray-detection in a new analytical electron-microscope, *J. Microsc.* **176**, 85-98 (1994).

51.   C. E. Lyman, Digital x-ray imaging of small particles, *Ultramicroscopy* **20**, 119-124 (1986).

52.   P. E. Batson, Simultaneous STEM imaging and electron-energy-loss spectroscopy with atomic column sensitivity, *Nature* **366**, 727-728 (1993).

53.   N. D. Browning, D. J. Wallis, P. D. Nellist and S. J. Pennycook, EELS in the STEM: Determination of materials properties on the atomic scale, *Micron* **28**, 333-348 (1997).

'54. J. Silcox, Core-loss EELS, *Curr. Opin. Solid St. M* **43**, 336-342 (1998).

55.   D. A. Muller and M. J. Mills, Electron microscopy: probing the atomic structure and chemistry of grain boundaries, interfaces, and defects, *Mater. Sci. Eng. A* **260**, 12-28 (1999).

56.   R. F. Klie, M. M. Disko, and N. D. Browning, Atomic scale observations of the chemistry at the metal-oxide interface in heterogeneous catalysts, *J. Catal.* **205**, 1-6 (2002).

57.   K. Sun, J. Liu, N. K. Nag, and N. D. Browning, Studying the metal-support interaction in Pd/γ-Al$_2$O$_3$ catalysts by atomic-resolution electron energy-loss spectroscopy, *Catal. Lett.* **84**, 193-199 (2002).

58.   J. Liu and N. K. Nag, Atomic resolution electron spectroscopy investigation of supported catalysts: Pd/TiO$_2$ and Pd-Ni/TiO$_2$, In *Proceedings of the 18th North American Catalysis Society Meeting.* pp 225-226 (2003).

59.   J. M. Cowley, Electron diffraction phenomena observed with a high resolution STEM instrument, *J. Electron Microsc. Tech.* **3**, 25-44 (1986).

60.   J. M. Cowley, Configured detectors for STEM imaging of thin specimens, *Ultramicroscopy* **49**, 4-13 (1993).

61.   J. M. Cowley, Electron nanodiffraction, *J. Electron Microsc. Tech.* **46**, 75-97 (1999).

62.   J. C. H. Spence and J. M. Zuo, *Electron Microdiffraction*, Plenum, New York (1992).

63.   J. M. Cowley and J. C. H. Spence, Convergent beam electron microdiffraction from small crystals, *Ultramicroscopy* **6**, 359-366 (1981).

64.   M. Pan, J. M. Cowley and J. C. Berry, Coherent electron microdiffraction from small metal particles, *Ultramicroscopy* **30**, 385-394 (1989).

65.   J. Liu, G. Hembree, G. Spinnler and J. Venables, High-resolution Auger electron spectroscopy and microscopy of a supported metal catalyst, *Surf. Sci.* **262**, L111-117 (1992).

66.   J. Liu, G. Hembree, G. Spinnler and J. Venables, Nanometer-resolution surface analysis with Auger electrons, *Ultramicroscopy* **52**, 369-376 (1993).

67   G. G. Hembree and J. A. Venables, Nanometer-resolution scanning Auger electron microscopy, *Ultramicroscopy* **47**, 109-20 (1992).

68.   P. Kruit, and J. A. Venables, High-spatial-resolution surface-sensitive electron spectroscopy using a magnetic parallelize, *Ultramicroscopy* **25**, 183-193 (1988).

69.   D. C. Joy and J. B. Pawley, High-resolution scanning electron microscopy, *Ultramicroscopy* **47**, 80-100 (1992).

70.   D. C. Joy and C. S. Joy, Low voltage scanning electron microscopy, *Micron* **27**, 247-263 (1996).

71.   J. Liu and J. M. Cowley, High-resolution scanning electron microscopy of surface reactions, *Ultramicroscopy* **23**, 463-472 (1987).

72.   D. J. Smith, M. H. Yao, L. F. Allard and A. K. Datye, High-resolution scanning electron microscopy for the characterization of supported catalysts, *Catal. Lett.* **31**, 57-64 (1995).

73.   R. Darji and A. Howie, Scattering corrections in small particle imaging, *Micron* **28**, 95-100 (1997).

74.   J. Liu, Low voltage high-resolution secondary electron microscopy of industrial supported catalysts, in: *Proceedings of the 14th ICEM: Electron Microscopy, Vol. 2*, H. A. C. Benavides and M. J. Yacaman, eds., Institute of Physics Publishing, Bristol, pp 399-400 (1998).

75.   J. Liu, Contrast of highly dispersed metal nanoparticles in high-resolution secondary electron and backscattered electron images of supported metal catalysts, *Microsc. Microanal.* **6**, 388-399 (2000).

76.   J. Liu, R. L. Ornberg and J. R. Ebner, Studies of supported metal catalysts using low voltage biased secondary electron imaging in a JSM-6320F FE-SEM, *Microsc. Microanal.* **3** (Suppl. 2), pp1123-1124 (1997).

77.   E. D. Boyes, High-resolution and low-voltage SEM imaging and chemical microanalysis, *Adv. Mater.* **10**, 1277-1280 (1998).

78.   J. Liu, High-resolution and low-voltage FE-SEM imaging and microanalysis in materials characterization, *Mater. Charact.* **44**, 353-363 (2000).
79.   J. F. Haw, *In Situ Spectroscopy in Heterogeneous Catalysis*, Wiley-VCH, Weinheim (2002).
80.   P. L. Gai, Direct probing of gas molecule-solid catalyst interactions on the atomic scale, *Adv. Mater.* **10**, 1259-1263 (1998).
81.   P. L. Gai, Probing selective oxidation catalysis under reaction conditions by atomic scale environmental high resolution electron microscopy, *Curr. Opin. Solid St. M* **4**, 63-73(1999).
82.   P. L. Gai, Developments in in situ Environmental Cell High-Resolution Electron Microscopy and Applications to Catalysis, *Top. Catal.* **21**, 161-173 (2002).
83.   P. L. Gai, Development of wet environmental TEM (wet-ETEM) for in situ studies of liquid-catalyst reactions on the nanoscale, *Microsc. Microanal.* **8**, 21-28 (2002).
84.   V. P. Oleshko, P. A. Crozier, R. D. Cantrell and A. D. Westwood, In situ real-time environmental TEM of gas phase Ziegler-Natta catalytic polymerization of propylene, *J. Electron Microsc.* **51**, S27-S39 (2001).
85.   P. A. Crozier, In situ nanoscale observation of polymer structures with electron microscopy, *Poly. Mater. Sci. and Eng.* **87**, 181-187 (2002).
86.   T. W. Hansen, J. B. Wagner, P. L. Hansen, S. Dahl, H. Topsoe and C. J. H. Jacobsen, Atomic-resolution in situ transmission electron microscopy of a promoter of a heterogeneous catalyst, *Science* **294**, 1508-1510 (2001).
87.   P. L. Hansen, J. B. Wagner, S. Helveg, J. R. Rostrup-Nielsen, B. S. Clausen and H. Topsoe, Atom-resolved imaging of dynamic shape changes in supported copper nanocrystals, *Science* **295**, 2053-2055 (2002).
88.   O. Scherzer, The theoretically attainable resolving power of the electron microscope, *Z. Physik* **114**, 427-434 (1939).
89.   O. Scherzer, Spherical and chromatic correction of electron lenses, *Optik* **2**, 114-132 (1947).
90.   J. Zach and M. Haider, Correction of spherical and chromatic aberration in a low-voltage SEM, *Optik* **98**, 112-118 (1995).
91.   M. Haider, G. Braunshausen, and E. Schwan, Correction of the spherical-aberration of a 200-KV TEM by means of a hexapole-corrector, *Optik* **99**, 167-179 (1995).
92.   M. Haider, H. Rose, S. Uhlemann, E. Schwan, B. Kabius, and K. Urban, A spherical-aberration-corrected 200 kV transmission electron microscope, *Ultramicroscopy* **75**, 53-60 (1998).
93.   O. L. Krivanek, N. Dellby and A. R. Lupini, towards sub-Angstrom electron beams, *Ultramicroscopy* **78**, 1-11(1999).
94.   N. Dellby, O. L. Krivanek, P. D. Nellist, P. E. Batson and A. R. Lupini, Progress in aberration-corrected scanning transmission electron microscopy, *J. Electron Microsc.* **50**, 177-185 (2001).
95.   H. W. Mook and P. Kruit, Optimization of the short field monochromator configuration for a high brightness electron source, *Optik* **111**, 339-346 (2000).
96.   H. W. Mook and P. Kruit, Construction and characterization of the fringe field monochromator for a field emission gun, *Ultramicroscopy* **81**, 129-139 (2000).
97.   A. J. Koster, R. Grimm, D. Typke, R. Hegrel, A. Stoschek, J. Walz and W. Baumeister, Perspectives of molecular and cellular electron tomography, *J. Struct. Biol.* **120**, 276-308 (1997).
98.   M. Weyland, P. A. Midgley, J. M. Thomas, Electron tomography of nanoparticle catalysts on porous supports: A new technique based on Rutherford scattering, *J. Phys. Chem.* **B105**, 7882-7886 (2001).
99.   P. A. Midgley, M. Weyland, J. M. Thomas, P. L. Gai and E. D. Boyes, Probing the spatial distribution and morphology of supported nanoparticles using Rutherford-scattered electron imaging, *Angew Chem Int Ed* **41**, 3804-3807 (2002).
100. M. Weyland, Electron tomography of catalysts, *Topics in Catalysis* **21**, 175-183 (2002).
101. G. Mobus, B. J. Inkson, Element specific nanoscale electron tomography, *Instit. Phys. Confer. Ser.* **168**, 267-270 (2001).
102. D. Gabor, A new microscopic principle, *Nature* **161**,777-780 (1948).
103. J. M. Cowley, Twenty forms of electron holography, *Ultramicroscopy* **41**, 335-348 (1992).
104. A. Tonomura, *Electron Holography*, Springer-Verlag, New York (1993).
105. A. Tonomura, L. F. Allard, G. Pozzi, D. C. Joy and Y. A. Ono, Eds., *Electron Holography*, Elsevier Science B. V., Amsterdam (1995).
106. M. Lehmann, H. Lichte, D. Geiger, G. Lang, E. Schweda, Electron holography at atomic dimensions - present state, *Mater. Character.* **42**, 249-263 (1999).
107. E. Volkl, L. F. Allard and D. C. Joy, Eds., *Introduction to Electron Holography,* Kluwer Academics, New York (1999).
108. L. F. Allard, E. Voelkl, D. S. Kalakkad and A. K. Datye, Electron holography reveals the internal structure of palladium nano-particles, *J. Mater. Sci.* **29**, 5612-5614 (1994).

109.   L. F. Allard, E. Voelkl, A. Carim, A. K. Datye and R. Ruoff, Morphology and crystallography of nanoparticulates revealed by electron holography, *Nanostruct. Mater.* **7**, 137-146 (1996).

110.   A. K. Datye, D. S. Kalakkad, E. Voelkl and L. F. Allard, Electron holography of heterogeneous catalysts, in: Electron Holography, eds., Tonomura, A., Allard, L. F., Pozzi, G., Joy, D. C., and Ono, Y. A., Elsevier Science B. V., Amsterdam (1995).

111.   M. Haruta, Size- and support-dependency in the catalysis of gold, *Catal. Today* **36**, 153-166 (1997).

19

Nano-Scale Characterisation of Supported Phases in Catalytic Materials by High Resolution Transmission Electron Microscopy

R. T. Baker, S. Bernal, J. J. Calvino, J. A. Pérez-Omil, C. López-Cartes.*

19.1. INTRODUCTION

In order to understand the macroscopic performance of a heterogeneous catalyst system, it is necessary to carefully consider the texture, morphology, phase composition, chemical composition and the relative dispositions and orientations, of all of the active and support phases of the catalyst - and not only the nature of the chemical reactions taking place at them - down to the nanometre and atomic scales. Changes in these properties induced by chemical and thermal treatments are also of great importance. High Resolution Transmission Electron Microscopy (HRTEM), and the family of related analytical techniques grouped around it, provide a very powerful way of studying catalyst materials down to precisely this scale.[1,2,3,4] Minute regions of a catalyst material can be visualised routinely in HRTEM images at resolutions of 2 Å or better. The crystallographic phase of nano-scale features can be determined by recording electron diffraction patterns in the microscope and, with the necessary auxiliary instruments attached, chemical compositional information at sub-nanometre resolution is also accessible. Recent and upcoming improvements in commercially available instruments,

* R. T. Baker, Division of Physical and Inorganic Chemistry, University of Dundee, Dundee DD1 4HN, United Kingdom. S. Bernal, J. J. Calvino, J. A. Pérez-Omil, Departamento de Ciencia de los Materiales e Ingeniería Metalúrgica y Química Inorgánica, Facultad de Ciencias, Universidad de Cádiz, Campus Rio San Pedro, Apartado 40, Puerto Real, 11510-Cádiz, Spain. C. López-Cartes, Instituto de Ciencia de Materiales de Sevilla (CSIC),Avda. Américo Vespucio s/n, Isla de la Cartuja, 41092 Sevilla, Spain.

such as the use of Field Emission Guns (FEGs) to generate highly coherent and intense electron beams and the incorporation of aberration correctors (C_s correctors), hold out the promise of increased precision and accuracy, access to more crystallographic and compositional information, applications to new areas, such as *in situ* studies,[5] and improved ease of use.

HRTEM is particularly powerful when combined with Image Simulation (IS).[6] Here, specialised software packages are employed to generate crystallographic model structures and to simulate the images they would give rise to in the microscope. These are then compared with experimental images. Valuable information can also be obtained by performing complimentary studies where HRTEM is used in parallel with techniques capable of probing the active sites of a specific catalyst system. These include Temperature Programmed Techniques, the chemisorption of probe molecules[7] and X-ray Photo-electron (XPS), infrared (IR), Nuclear Magnetic Resonance (NMR)[8] and other spectroscopies.

In this contribution, we will illustrate how different classes of information relevant to catalysis can be obtained using HRTEM techniques by referring to a number of studies performed on a range of heterogeneous catalyst systems. These studies will also clearly exemplify the value of using IS and the utility of complimentary studies. Catalyst systems investigated include supported noble metal nano-particles as well as supported oxide structures.

19.2. HIGH RESOLUTION TRANSMISSION ELECTRON MICROSCOPY

19.2.1. The Transmission Electron Microscope

The electron microscope was initially conceived as a response to the limitations of light microscopes for imaging very small objects shortly after de Broglie's discovery that electrons could be treated theoretically as waves. The resolution of a microscope is the minimum distance at which two objects may be distinguished and its theoretical maximum is proportional to the wavelength of the radiation used. Hence, the maximum theoretical resolution of a light microscope is of the order of a few hundred nanometres (nm). The wavelength of an electron is related to its energy such that an electron accelerated to 100 kilo-electron Volts (keV) will have a wavelength, and a maximum theoretical resolution, of the order of a few picometres (pm),[9] about 100 times smaller than the diameter of an atom. The potential of the electron microscope to attain such sub-atomic resolution has not yet been fulfilled because of the considerable practical difficulties involved in microscope construction. However, atomic-scale resolution is attainable using modern instruments.

As well as this imaging capability, other important advantages of using electron beams were soon recognised. Electrons interact strongly with atoms and can be employed in crystallographic diffraction experiments in the Transmission Electron Microscope (TEM).[10] Unlike X-rays, electrons are charged and can be directed and focused easily using magnetic and electric fields. This allows electron diffraction to be performed down to the nano-scale to extract crystallographic information from tiny areas of a sample, such as a single crystallite or part of a crystallite. The resulting Selected Area Electron Diffraction (SAED) patterns consist of a 2-D array of "spots" for a single crystal which

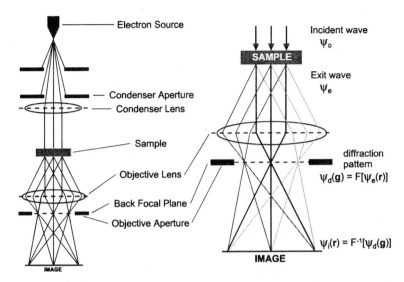

FIGURE 19.1. Simplified Schematic Ray Diagrams of the HRTEM in Imaging Mode Representing the Functions of Electron Source and Condenser and Objective Lenses. On the Right, the Relationship between the Exit Wave, the Diffraction Pattern Formed at the Back Focal Plane and the Image are Represented.

can be related to a slice through the reciprocal lattice of the diffracting crystal.[11] Inter-planar distances and the angles between planes are thus obtained. If many crystals are selected, the angular information is lost and the pattern resembles a series of concentric rings from which inter-planar distances only can be calculated.

The Transition Electron Microscope consists of a vertical column down which the electron beam passes from the electron source at the top, through the specimen and down to the bottom of the column where the image is formed. Figure 19.1 contains a simplified representation of the TEM. The column is held under vacuum by a system of high performance pumps to reduce scattering of the electron beam by gas atoms. Electromagnetic coils, known as lenses, are positioned around the column along its length and are used to modify the motion of the electrons in the beam, either concentrating or spreading the beam as required, in an analogous way to the optical lenses in a light microscope. Apertures of different diameters can be inserted into the electron beam at several positions along the column. This is done to select part of the beam, for image formation, for example, and exclude the contribution of the rest.

The electron beam is generated in the electron gun at the top of the microscope column. Free electrons are obtained from a filament in one of two ways. Still the most common arrangement involves the thermionic emission of electrons from a filament of tungsten or LaB_6. These filaments are heated by an electric current to about 2500°C and 1500°C, respectively, at which temperatures a significant number of electrons obtain energy equal to or greater than the work function of the material and are able to leave the filament material and pass into the vacuum. These emergent electrons are focused by a cylinder to which a small negative bias is applied. At the same time, they are accelerated by a very large electrostatic field applied between the filament and the anode plate and they pass through an aperture in the anode plate, then into the TEM column. The strength of this field determines the kinetic energy of the electron beam and therefore the wavelength of the electrons and their maximum theoretical resolving power (see above). In practice, an applied potential of at least 100 kilovolts (kV) is advisable for HRTEM. The electron beam may also be generated in a Field Emission Gun (FEG). These work at much lower temperatures, about 300°C, and contain two successive electrostatic fields, between the filament and two anode plates positioned one after the other. The tungsten filament in a FEG has an extremely fine tip. The first electrostatic potential extracts electrons from the filament through this tip. The second potential is used, as above, to accelerate the electrons and form the high energy beam. Because the filament tip is so fine, the beam from a FEG is much more intense and much more spatially coherent than the beam from a thermionic filament. Furthermore, the variation in energy within the beam is very small. These factors make FEGs of great interest for HRTEM where a narrow, coherent and intense beam is required.

The electron gun generates a thin electron beam by focusing and accelerating electrons from the filament. Electrons with energies or paths different from those required are excluded from the beam at the apertures. The Condenser Lens system is used to further adjust the beam so that it has the desired properties when it arrives at the specimen. Usually, a parallel beam is required. Different beam diameters may be desired and these can be set using the Condenser Lenses (there are usually two or three). This beam then passes through the specimen which is mounted on the specimen holder. The incident beam (see Figure 19.1), interacts with the internal crystal structure of the sample and emerges as a set of diffracted and non-diffracted beams. These beams are focused by the Objective Lens. In Figure 19.1, a ray diagram is drawn containing the non-diffracted beam and two diffracted beams. These are focused to points, or spots, at the Back Focal Plane (BFP). We see that there is no direct relationship between the position of these spots and the point of origin of the beam in the sample. However, it is clear that beams diffracted at the same angle are focused to the same spot in the BFP. Considering the rule for Bragg diffraction, and extending to three dimensions, we would expect each spot to be related to one particular value of inter-planar spacing at one particular orientation - that is, to one family of crystal planes at one particular orientation. This is what we would expect to see in a diffraction pattern and that is indeed what is generated at the BFP under these conditions. Following the three rays further, we see that they are again focused at the image plane. Here, all the rays from each individual spot can be traced back to the same point in the sample, confirming that the image of the sample is formed at this plane. In fact, the image and the diffraction pattern formed by a particular crystal in a particular orientation are related mathematically, the latter being the Fourier Transform of the former. This is illustrated in Figure 19.1 where the incident wave, ψ_o, exit wave, ψ_e, and the wave functions of the diffraction pattern, ψ_d, and the image, ψ_I are labelled. The incident wave, ψ_o, interacts with the internal potential of the sample and emerges as ψ_e, which therefore contains information from the sample. The Objective Lens effectively performs a Fourier Transform on ψ_e to form the diffraction pattern which is a Fourier Transform, F (in reciprocal space, denoted by the vector, $\mathbf{g}$), of the exit wave (in real space, denoted by the vector, $\mathbf{r}$). The image is an inverse Fourier Transform, F^{-1} (in real space), of the diffraction pattern and is therefore closely related to the internal structure of the sample. A very widely used technique in the analysis of HRTEM images is to perform a Fourier Transform mathematically on the area of interest of an image in order to calculate a Digital Diffraction Pattern, or DDP. Information about the crystallography of the lattice images and the amplitude and phase of the Fourier components of the image (lattice planes), can be easily extracted. This in turn allows the inter-planar spacings and inter-planar angles to be obtained from the selected area of the image. The use of DDPs is often more convenient and more precise than collecting a SAED pattern experimentally, especially when examining very small regions.

Images and diffraction patterns are visualised on a fluorescent screen and may be recorded either on photographic film, or, increasingly, digitally, using a camera positioned below the fluorescent screen. To record a diffraction pattern, the strength of the Objective Lens is changed in order to throw the diffraction pattern onto the image plane. Further electromagnetic lenses are incorporated between the BFG and the image plane in commercial TEM instruments to allow the operator to focus and adjust the magnification of the final image. For clarity, these are not included in Figure 19.1. Several books are available on Electron Microscopy and HRTEM which provide more depth than is possible here.[12,13,14]

19.2.2. Related Quantitative Techniques

As well as the elastic interactions of electrons with the specimen which give rise to the image and the diffraction pattern, their inelastic interactions, in which energy is transferred, also provide useful complimentary information and powerful analytical techniques have grown up around the TEM to exploit this.[15] An incoming high energy electron may transfer energy to one of the core electrons of an atom such that it is ejected from the atom. An electron at a higher energy level in the atom may then fall into the resulting electron hole releasing its excess energy as an X-ray photon. The frequency of such a photon is determined by the difference between the two electron energy levels and will therefore be characteristic of the nature of the atom from which it was emitted. In X-ray Energy Dispersive Spectroscopy (XEDS or EDX),[15] these X-ray photons are collected and their energy plotted against number of counts to give a spectrum. This technique allows quantitative measurement of the elemental composition of areas of the sample, again, down to the nano-scale. In Electron Energy Loss Spectroscopy (EELS),[15] the electrons themselves are collected, after losing energy to the sample. The amount of energy lost can be measured accurately and can be related back to the differences between electronic energy levels in the atoms of the specimen. Information on the oxidation states and chemical environment of these atoms, at a resolution of a few Ångstroms, can be extracted from the EELS spectra. Again, this technique provides information which is not available from HRTEM in itself.[16] For this reason, both XEDS and EELS spectrometers are commonly fitted to the modern TEM. The application of these and other related quantitative techniques is described in the previous chapter.

19.2.3. Limitations of HRTEM

HRTEM has several limitations[1] which must be remembered. Since it is a transmission technique, the samples must be electron transparent and, in addition, must be thin enough to minimise multiple diffraction of the electrons. In practice, specimens must be less than about 100 nm thick, although this is dependant on the composition of the material. In heterogeneous catalysis, where specific surface areas and particle dispersions are usually high, this is often not a serious problem. The high energy electron beam may alter and damage the specimen by heating it or by reducing susceptible materials such as some oxides. These difficulties can be ameliorated somewhat by the use of modern digital cameras which are so sensitive that a spread beam can be used in order to expose the sample to less intense radiation and good results still be obtained. Interpretation of HRTEM images requires great care since they are a 2-D representation of a 3-D set of objects viewed by transmission rather than reflection. The image may contain several superposed crystals, for example, and therefore will contain information from multiply diffracted electrons. In a simple illustration of this, Figure 19.2 is an image of a catalyst sample in which Pt particles are dispersed over a CeO_2 support. The Pt particles are imaged in two modes: in profile view (black arrows) where the electron beam passes only through the Pt particle and therefore carries information relating to it only; in plane view (white arrows), where the electron beam passes through the support material as well as through the Pt particle and will therefore carry information relating to both.[17] Particles imaged in profile view are, therefore, more easily interpreted. Those imaged in plane view are generally darker and, importantly, exhibit broad fringes caused by the double diffraction of the electron beam by the support material and the metal particle. These are known as Moiré fringes.[18]

Unless the HRTEM sample is extremely thin, the strength of the electron-matter interaction means that the diffraction process is usually dynamic. That is, scattering is multiple instead of single, as is usual in the case of X-Rays. Under dynamic diffraction conditions the amplitude and phase of the unscattered and diffracted beams deviate from those expected from structure factors (F_{hkl}). In general, this means that a direct correlation does not exist between the contrasts seen in the image and the projected structure of the material. The non-linear characteristics of the imaging process, which in turn is strongly influenced by the operating conditions of the microscope, also contribute to degrade the correlation between the real structure and the projected structure observed in the HRTEM image. The aberrations of the lens systems as well as the particular recording conditions of the image affect this correlation. Therefore, only in particular diffraction and image recording conditions can the image be considered to be a structural image, showing the projection of atomic columns in the material at their correct positions. The more general case is that the image is what is called a lattice image. In this case the spacings and angles of the fringes recorded on the image are those present in the material, but the exact position of black and white contrasts do not resemble the positions of the atomic columns. In other words, the crystallographic information concerning the atomic planes is correct but a direct retrieval of the atomic positions is not possible. Nevertheless, such crystallographic information is usually rich enough to provide valuable structural and phase composition data for catalyst samples.

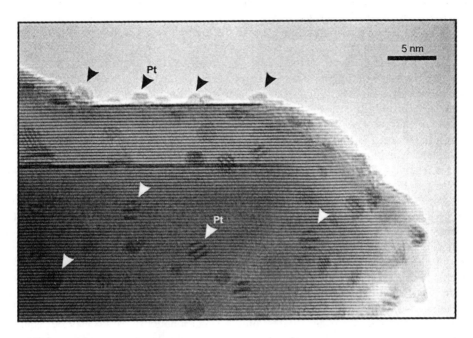

FIGURE 19.2. HRTEM Image of Nano-Scale Pt Particles Supported on CeO_2 in a 4 wt.% Pt/CeO_2 Catalyst Sample after Reduction at 350°C. Particles Are Seen in Both Profile View (Black Arrows) and Plane View (White Arrows).

19.2.4. Preparation and Visualisation of Catalyst Samples

The majority of catalyst samples require only relatively simple preparation for study in HRTEM since they are fine powders. There is, therefore, no need to thin the samples since they are already electron-transparent. Usually a copper or gold grid covered with a layer of carbon film with either a lacy or holey texture is used. These are commercially available. The catalyst powder is suspended in a volatile liquid such as an alcohol, acetone, or an alkane such as hexane, preferably by ultrasonication, and a drop of the suspension placed on the grid. Alternatively, the grid may be passed through the suspension using tweezers. Either process may need to be repeated several times, with time allowed for evaporation of the liquid, to obtain a sufficient loading of catalyst on the grid. The particles of the catalyst adhere to the carbon and those which overhang holes in the structure may be imaged in transmission. Catalyst loading of the grid should be verified using an optical microscope before use in the TEM.

In the technique of microtomy the material of interest is sliced into layers a few tens of nanometres thick using a quartz or diamond blade mounted on a mechanical arm, a microtome. This preparation technique is used extensively for the investigation of biological samples held in solidified resin. It is also useful for studying catalyst powder samples in resin. One advantage is that overlap of particles may be reduced since the thickness of the sample is limited by the thickness of the section made. It may then be easier to visualise the crystal structure of the particles, especially if they tend to form agglomerates. In addition, since particles or agglomerates may be broken open, differences between the surface and the the bulk structure and composition may be imaged. This may be of interest to determine the distribution of supported metal particles or the nature of an encapsulating surface layer.

To study thick, dense objects such as semiconductor wafers or ceramics, for example, ion beam milling must be employed to reduce the thickness of the sample in the area of interest. A hole is milled from one side of the sample and the thin region around the hole is examined in the TEM since the edge of the hole will be thin enough to transmit electrons. This is a time-consuming process and requires specialist equipment. However, where it is important to visualise catalysts supported in pellets after use, for example, this method may provide more information about the disposition of the active phases relative to the support material than would a study of the powder obtained from grinding up the pellet.

In order to visualise the crystal structure of a particle in HRTEM, the particle must be orientated in such a way that its atom columns are aligned with the electron beam. That is, a zone axis of the crystal structure must be aligned with the electron beam. A zone axis describes a direction in the crystal which is parallel with at least one family of crystal planes. Most electron microscope sample holders have the facility for tilting the sample in at least one direction to allow the operator to align particles with the electron beam. The maximum angle through which the sample can be tilted is limited by the design of the sample holder and by the internal geometry of the TEM instrument. In the case of a fine catalyst powder sample, only a small amount of tilting may be required since, because of the large number of particles, it may be more profitable simply to search for particles which are already close to alignment. Care should be taken, however, since the smaller the particle, the further it can be from true alignment and still show its interatomic planes. One technique of special relevance to the study of catalysts is that of surface profile imaging.[19,20] Here a particle is selected and tilted until one of its surface planes as well as a set of atom columns are parallel to the electron beam. Under well-controlled operating conditions, this allows the imaging of surface features which are very important in heterogeneous catalysis such as surface steps, restructured surfaces, particle decoration, missing atomic rows and other defects.

19.2.5. Image Simulation

Because of the complexity of HRTEM image interpretation, for the detailed analysis of the contrasts in HRTEM images and to model accurately the diffraction and image formation processes, Image Simulation (IS) techniques are necessary.[6] The procedure for Image Simulation can be broken down into three steps. Firstly, a mathematical representation of the structure observed in the experimental image must be constructed on Cartesian co-ordinates. Sophisticated software packages are available for the generation of these structural computer models. Many variables must be considered in this step. An example relevant to heterogeneous catalysis is that of a metal particle supported on an oxide support. The crystallographic phases of the particle and the support must first be decided. There may be several possible alternatives. For example, tetragonal and cubic structures of the oxide may both be known. The relative orientation of the crystal

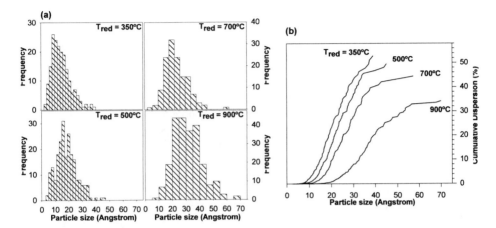

FIGURE 19.3. (a) Rh Particle Size Distributions and (b) Calculated Cumulative Dispersion Plots for Samples of 0.78 wt.% Rh/Ce$_{0.68}$Zr$_{0.32}$O$_2$ Reduced for 1 h at Reduction Temperatures (T$_{red}$) of 350, 500, 700 and 900°C. Reprinted with Permission from Reference 21. Copyright (2000) American Chemical Society.

structures of the particle and the support, as well as the nature of the interface where the two meet, must be decided. For example, is the particle epitaxial with the support, and if so which planes are parallel in the two structures? The thickness of the support crystal and the size of the supported particle are set by inputting the number of crystal planes in each direction. The morphology of the supported particle is set by adding the facetting of the particle, *i.e.* by identifying the crystallographic planes which make up the particle's surface. Close inspection of the experimental images and a thorough knowledge of the crystallography of the phases which may be present in the sample are essential in setting these variables. Even so, some uncertainty will remain in some of them. To accommodate this, and to follow good practice, several alternative structural models are generally prepared for any particular experimentally-observed feature. These are then passed on to the second step of the IS process in which the interaction of these candidate structures with the electron beam of the TEM is modelled. Each structural model, its alignment with respect to the electron beam and the operating parameters of the microscope used to obtain the experimental images are input into the computer. The effects of the interactions of the electron beam with the atoms of the model structure, and with the apertures and electromagnetic lenses of the TEM itself, are modelled mathematically and the images which would result are simulated. Important variables in this step are the exact geometry of the alignment of the model structure to the electron beam and the defocus value – a measure of the extent to which the specimen is out of focus. Both of these variables can cause dramatic changes in the contrasts seen in the simulated images. Because of this, it is normal practice to collect both experimental and simulated images over a range of defocus values - a through-focus series. In the final step of the IS process, the series of images simulated for each model structure are compared with the experimental images and those proposed structures giving a poor match are excluded, leaving the more probable alternatives for further study.

19.3. PARTICLE SIZE ANALYSIS

In catalyst materials containing phases dispersed over a support, the size, size distribution and dispersion of the supported particles, and often of the support particles, is of great importance, as is the evolution of these parameters during thermal and chemical treatments. By recording numerous HRTEM images and then identifying and measuring the size of the particles of interest, such information can be obtained. In order for the data to be representative and statistically meaningful, many images of many different regions of each sample must be obtained and several hundred particles must be measured. The measurements themselves may be made adequately by hand from micrographs. However, it is more convenient and more accurate to perform such measurements from digital images with the aid of appropriate software.

Particle size distributions are plotted in Figure 19.3(a) for Rh particles dispersed over the surface of the mixed oxide support in a 0.78 wt.% $Rh/Ce_{0.68}Zr_{0.32}O_2$ automotive catalyst which was the subject of a recent study.[21] Samples of the catalyst were reduced in H_2 for 1 h at different reduction temperatures, T_{red}, as indicated on the Figure. Sample size and mass-corrected mean particle size data are summarised in Table 1. The mass-correction is applied in order to weight the data in terms of the mass of particles of a particular size range, rather than simply in terms of the number of particles of that size range, as shown in Figure 19.3(a).

A procedure which is particularly useful to the study of heterogeneous catalyst materials is the ability to obtain dispersion data from the particle size distributions. Here, dispersion is defined as the ratio of total exposed particle surface area to total particle mass or, equally, as the total number of metal atoms at an exposed surface divided by the total number of metal atoms. The relevant mathematical treatments are given and are discussed in detail elsewhere.[22,23] Briefly, a morphology and an orientation with respect

TABLE 19.1. Comparison of Rh Particle Dispersion Data from Volumetric H_2 Chemisorption and from HRTEM Measurements for $Rh/Ce_{0.68}Zr_{0.32}O_2$. Reprinted with Permission from Reference 21. Copyright (2000) American Chemical Society.

Run[a]	Reduction temperature (°C)	by HRTEM			by chemisorption	
		Sample size	Mean particle size[b] (Å)	Rh_S/Rh_T	H/Rh_T (-80°C)	H/Rh_T (25°C)
1	150[c]	231	24	0.52	0.52	2.38
2	350	189	25	0.53	0.57	0.68
3	500	221	27	0.49	0.49	0.47
4	700	135	32	0.44	0.41	0.38
5	150[d]				0.47	
6	900	234	42	0.34	0.23	0.20
7	150[d]				0.31	

[a] catalyst evacuated for 4 h at 400°C before chemisorption measurement
[b] mass corrected
[c] data not included in Figure 19.3
[d] catalyst from previous run, re-oxidised at 427°C and reduced at 150°C

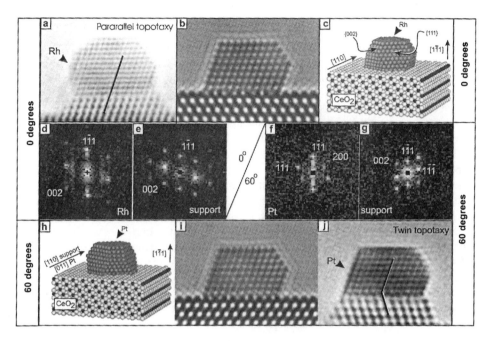

FIGURE 19.4. (a) Experimental HRTEM Image of a Supported Rh Particle in a 0.5 wt.% $Rh/Ce_{0.8}Tb_{0.2}O_{2-x}$ Catalyst Sample; (b) Corresponding Simulated Image; (c) Structural Model Used to Generate Simulated Image; (d, e) DDPs Obtained from the Experimental Image for the Metal and the Support, Respectively; (j) Experimental Image of a Supported Pt Particle in a 4 wt.% Pt/CeO_2 Catalyst; (i) Corresponding Simulated Image; (h) Structural Model Used to Generate Simulated Image; (f, h) DDPs Obtained from the Experimental Image for the Metal and the Support, Respectively.

to the surface of the support are assumed for the particles. For the Rh particles studied here, a truncated cuboctahedral morphology was assumed with {111}-type planes of both the metal and the oxide support in contact at their shared interface (assuming cubic crystal structures for both).These assumptions were based on detailed analysis of high resolution images of several representative particles. Based on this morphology, the exposed surface area and volume of each measured particle is estimated. Hence the contribution to the overall dispersion value of the sample made by each particle is also estimated and plotted against particle size in a graph of cumulative dispersion, a set of which are presented in Figure 19.3(b) for the $Rh/Ce_{0.68}Zr_{0.32}O_2$ samples discussed here. The dispersion value for the sample is given by the sum of all contributions, $i.e.$ by the point on the extreme right of the curve, and these are given for $Rh/Ce_{0.68}Zr_{0.32}O_2$ in Table 1. It can now be seen that mean particle size increases and the particle size distribution broadens on increasing the reduction temperature. Dispersion, as estimated from HRTEM measurements, decreases from a value of 52% (or $Rh_S/Rh_T = 0.52$ where S and T subscripts refer to surface and total, respectively) for a reduction temperature of 150°C to 34% (0.34) for $T_{red} = 900°C$. In summary, the particles appear to undergo a classic evolution that can be explained by progressive sintering of the Rh particles.

In this study, it was interesting to compare the dispersion values estimated from the HRTEM measurements with those calculated from volumetric H_2 chemisorption measurements. Furthermore, the chemisorption was performed at two experimental temperatures, 25°C and -80°C. These data are compared in Table 1. Firstly, excellent agreement is achieved between the HRTEM values and the chemisorption values obtained at the -80°C for reduction temperatures of 150°C to 500°C. The chemisorption values obtained at 25°C are much higher, and higher than unity for $T_{red} = 150°C$. This suggested that an activated spillover process of hydrogen species onto the support occurred during the chemisorption measurements made at the higher temperature for the samples reduced at the lower reduction temperatures. Secondly, at $T_{red} > 500°C$, the chemisorption values (measured at -80°C) drop steadily below those obtained by HRTEM. However, after a mild re-oxidation and low temperature reduction, the higher dispersion values in the chemisorption experiment were recovered. This suggested a deactivation of the chemisorption of H_2 at the higher reduction temperatures which is not related to the sintering of the particles, since this is detected by HRTEM. Possible mechanisms proposed were partial coverage of the metal surface (see Section 19.5) by oxide material from the support or an electronic deactivation of the chemisorption process due to the influence of the reduced support on the band structure of the metal particle. Since little decoration was observed by HRTEM, and because reversal of the deactivation effect was possible under mild conditions which would be unlikely to reverse metal particle decoration, the second mechanism was accepted. This example illustrates the power of combining HRTEM with chemisorption techniques.

19.4. PARTICLE MORPHOLOGY AND RELATIONSHIP TO SUPPORT

19.4.1. Supported Metal Particles

The data in Figure 19.4 have been taken from various studies[24,25,26] and are a good illustration of the power of HRTEM when used in combination with Image Simulation and with the calculation of DDPs from high quality images. Experimental images of two supported metal particles are presented. In Figure 19.4(a), a Rh particle of approximate diameter 3 nm is supported on a mixed oxide of $Ce_{0.8}Tb_{0.2}O_{2-x}$.[24] The sample was reduced at 500°C in H_2 for 1 h. The crystallographic planes of the support material and the Rh metal particle are extremely well-defined. Such a high resolution image allows the accurate preparation of a composite structural model in which the orientation of the support and metal crystals to the electron beam and their relationship to each other can be estimated with a high degree of confidence. The morphology of the metal crystallite, in terms of the identities of its surface planes, is also included in the structural model. Typically, several different models would be constructed, each including variations in the parameters just discussed. An example of such a model, corresponding to the experimental image in Figure 19.4(a) is shown in Figure 19.4(c). The structural model is input into another software package which simulates the imaging process in the microscope. Both electron-optic parameters characteristic of the microscope and operational parameters, such as the defocus at which the image was recorded, are input and, by calculating the nature of the interaction of the electron beam with the structure represented in the structural model, HRTEM images are simulated. Such an image is

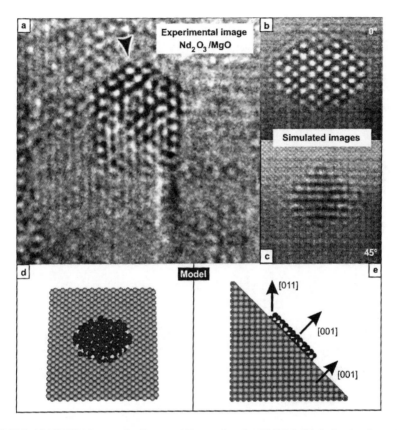

FIGURE 19.5. (a) HRTEM Image of a Supported Feature in a 3wt.% Nd₂O₃/MgO Catalyst Imaged in Plan View; (b, c) Corresponding Simulated Images; (d, e) Structural Model Constructed of the Feature Imaged in (a) in Plan and Profile Views, Respectively.

shown in Figure 19.4(b) and a strong general similarity with the experimental image is evident. It is clear in the experimental image that the crystallographic planes of the Rh metal are aligned with those of the support oxide material (see construction line). This is one of a set of general orientations found for supported metal nanoparticles after reduction treatments.[27] It is further illustrated by calculating DDPs for areas of the image containing only the Rh metal and only the support. Such DDPs are presented in Figures 4(d) and (e). Each spot represents a family of parallel planes of the same inter-planar spacing. The direction of the line from the centre spot (the equivalent to the undiffracted beam in a SAED) is normal to these planes in the image. Hence, the two major visible, exposed facets of the Rh particle can be identified as (1-11) and (00-2), for example, and the interface between the support and the particle appears also to be a (1-11) plane. The fact that the spots in the two DPPs are aligned in the same way implies that the crystal structures are indeed in the same crystallographic orientation. That is, all planes are aligned. This is known as the parallel orientation case. Although this can be seen clearly

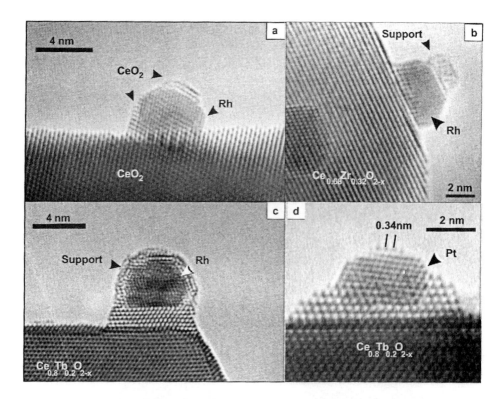

FIGURE 19.6. Decoration of Supported Metal Particles by Support Material. Supported Rh Particles in (a) 2.4 wt.% Rh/CeO$_2$ Reduced at 900°C; (b) 0.78 wt.% Rh/Ce$_{0.68}$Zr$_{0.32}$O$_{2-x}$ Reduced at 900°C; and (c) 0.5 wt.% Rh/Ce$_{0.8}$Tb$_{0.2}$O$_{2-x}$ Reduced at 900°C. (d) 5 wt.% Pt/Ce$_{0.8}$Tb$_{0.2}$O$_{2-x}$ Reduced at 700°C.

in the image in this example, DDPs are often very useful when the image contains more complex fringes; where two or more phases are superposed, for example.

Figure 19.4(f) to (j) present analogous information for the reduced Pt/CeO$_2$ catalyst.[25] Again, a structural model was constructed and the simulated image shows excellent agreement with the experimental image. The orientation of the fringes is well reproduced. In this example, the (1-11) planes of the metal and the CeO$_2$ crystals are aligned, as for the Rh particle. However, the interface between the particles can be considered as a mirror plane which reflects the (-1-11) plane of the Pt onto the (1-1-1) plane of the CeO$_2$. This gives rise to a rotation of the DDP, in which it can also be seen that only the (1-11) planes remain aligned. This case is known as a twin, or a 60° rotated, orientation. The relationship between this twin and the parallel orientation is that the metal particle is rotated by 60° with respect to the support around the [111] axis normal to the metal/support interface. The interfaces are crystallographically identical in both cases since the metal plane and oxide plane which are in contact both have C$_6$ symmetry.

19.4.2. Supported Oxide Phases

Since many catalyst systems consist of supported oxides, it is important that the structures of such systems can be interpreted by HRTEM. An excellent demonstration of the great sensitivity of HRTEM and IS is provided by the example given in Figure 19.5(a). This shows an image of a supported nano-scale feature observed in a 3 wt.% Nd_2O_3/MgO catalyst sample.[24] The sample was studied as prepared. Little information regarding the support can be extracted from the image. However, the supported feature (arrowed) is reasonably well defined and is made up of light dots which are spaced periodically in the two dimensions of the image. Several structural models were constructed and images were simulated for comparison with the experimental image. The model which gave the best match with the pattern in the experimental image was a rounded, monolayer thick raft of cubic Nd_2O_3 viewed in the [001] direction and supported on a (001) surface plane of the MgO support in parallel orientation. This model is shown in plan and profile view in Figure 19.5(d) and (e), respectively. The quality of the model is demonstrated by the close resemblance of the corresponding simulated image (Figure 19.5(b)) to the pattern observed experimentally in all characteristic details.

Interestingly, the relationship between the Nd_2O_3 raft and the MgO support at the interface is incoherent in that only every third Nd_2O_3 (400) plane coincides (approximately) with every fourth MgO (200) plane. There is a mismatch in this 3:4 correspondence of 1.6%. A second structural arrangement which was also modelled involved rotating the Nd_2O_3 raft by 45° with respect to the support around the common [001] direction, starting from the orientation described above. This would lead to a coherent 1:1 correspondence at the interface, but with a rather larger mismatch of 7.2%. The image simulated using this structural model (Figure 19.5(c)) is clearly very different from the first simulation and shares much fewer characteristics with the pattern in the experimental model. Therefore, the first model is the most suitable. The large difference in the simulated images caused by simply rotating the Nd_2O_3 raft by 45° shows how sensitive the HRTEM technique is, and how it can be used in conjunction with detailed analysis to identify and differentiate even such difficult structures as these.

19.5. INTERACTION OF THE SUPPORT WITH DISPERSED PARTICLES

The minimisation of catalyst deactivation, and the reactivation of used catalyst, are immensely important industrially and depend on a fundamental understanding of the mechanisms of deactivation and reactivation. For catalysts which contain supported metal particles on reducible supports, the term Strong Metal-Support Interaction (SMSI)[28,29,30] covers such deactivation mechanisms which proceed via an interaction between the support material and the dispersed metal phase, under reducing conditions. These include electronic effects, which decrease the metal's activity for chemisorption; the physical prevention of contact between the gas phase and the metal surface due to the growth of surface layers of support material (decoration); and even the formation of alloys with elements from the support. We have seen (Section 3) that the use of HRTEM to obtain particle size information and dispersion data in concert with other, particularly adsorption, techniques can aid the understanding of deactivation and SMSI mechanisms.

The direct imaging of supported metals by HRTEM is also very valuable for determining

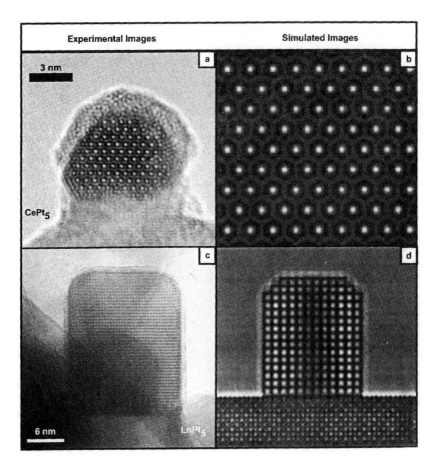

FIGURE 19.7. (a) Experimental HRTEM Image of an Intermetallic Phase Formed in a 4 wt.% Pt/CeO$_2$ Catalyst Reduced at 900°C; (b) Corresponding Calculated Image; (c) Experimental HRTEM Image of an Intermetallic Phase Formed in a 5 wt.% Pt/Ce$_{0.8}$Tb$_{0.2}$O$_{2-x}$ Catalyst Reduced at 900°C. (d) Corresponding Simulated Image.

the importance of nano-structural SMSI effects. One such effect is the decoration or encapsulation of the active dispersed metal phase by material from the support. There is strong evidence that several support oxides become mobile when they are partially reduced at elevated temperatures. Bernal *et al*[25] give a very detailed review of the preparation and characterisation of supported metal catalysts on ceria and ceria-based mixed oxide supports and of the application of HRTEM to the characterisation of SMSI mechanisms for a range of support oxide materials.[31]

Some examples of decorated particles observed in various catalysts are given in Figure 19.6. Rh-containing catalysts only appear to exhibit decoration after reduction at very high temperatures when supported on ceria or Ce-containing mixed oxides. In

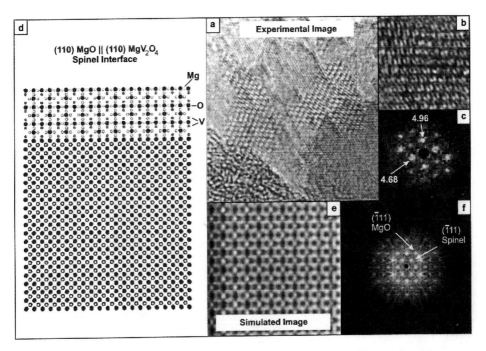

FIGURE 19.8. (a) Experimental HRTEM Image of Part of a 14%V/MgO Catalyst after Exposure to Pure Propane at 500°C for 4 h. (b) Detail of One of the Fringes of Interest. (c) DDP Calculated from the Experimental Image. (d) Structural Model Constructed to Explain the Experimental Contrasts. (e) Image Simulated Using this Structural Model. (f) DDP Calculated from the Simulated Image.

Figure 19.6(a) to (c) the reductions were performed in H_2 at 900°C for 1 h. As shown here for the CeO_2,[26] $Ce_{0.68}Zr_{0.32}O_{2-x}$[21] and $Ce_{0.8}Tb_{0.2}O_{2-x}$.[27] supports, the decorating layers are crystalline. The images show the presence of thin surface layers (Figure 19.6(a)), a thicker layer (Figure 19.6(b)) and the formation of a "pedestal" (Figure 19.6(c)) whereby the Rh particle appears to be raised above the level of the original support crystal surface by the growth of layers of support oxide below the metal particle. The alignment of the crystallographic planes of the Rh metal and the pedestal is shown clearly in this image. The Rh particle appears to be completely encapsulated by the support. The detection of small amounts of decoration is problematical unless the particle is perfectly focused. However, as an illustration of what can be achieved, Figure 19.6(d) shows a Pt particle which appears to have been in the process of being encapsulated. On the upper face however, there is only a small monolayer island of support material, containing approximately 30 atoms. Even so, it can reasonably be assigned as support material from its inter-planar spacing and by comparison of the experimental image with images simulated using a representative structural model.[32]

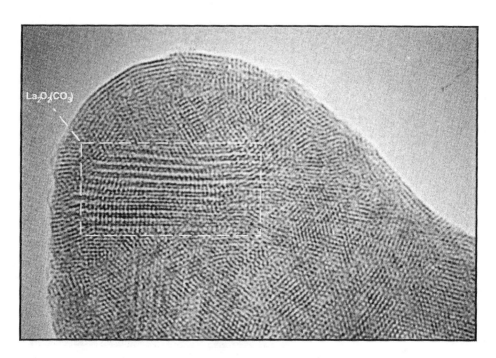

FIGURE 19.9. HRTEM Image of Polycrystalline Particle in a 10%Rh/La$_2$O$_3$ Catalyst Sample after Impregnation Using an Acidic Rh Precursor Solution, Drying and Reduction in H$_2$ at 200°C for 1 h.

19.6. STRUCTURAL AND PHASE CHANGES

19.6.1. Formation of Intermetallic Alloys

Chemical reaction of the support with the metal particles dispersed over its surface is one possible mechanism for catalyst deactivation via a SMSI. Pt particles supported on CeO$_2$[33] and Ce$_{0.8}$Tb$_{0.2}$O$_{2-x}$[34] have been found to form intermetallic alloys of Ce and Pt and Ce, Tb and Pt, respectively. These reactions have only been observed in severely reducing conditions and at very high temperatures, typically in H$_2$ at 900°C for 1 h. Furthermore, of the five known Ce-Pt intermetallic phases, only the hexagonal CePt$_5$ phase has been detected in these catalyst samples. A 6 nm supported particle with a core of CePt$_5$ is shown in Figure 19.7(a). The intermetallic phase is viewed along the [001] zone axis. The large lattice spacings and the hexagonal symmetry are inconsistent with all elemental phases of Pt and, amongst the intermetallics, only match the crystallography of the CePt$_5$ composition. An image calculated for this phase viewed along the same zone axis (Figure 19.7(b)) shows the excellent agreement with the experimental pattern.

When Tb is included with Ce in oxide support material, a new phase which is isostructural with CePt$_5$ is observed after very severe reduction treatments. A crystal of this phase is shown in Figure 19.7(c). It was assigned as LnPt$_5$ (where Ln may be Ce

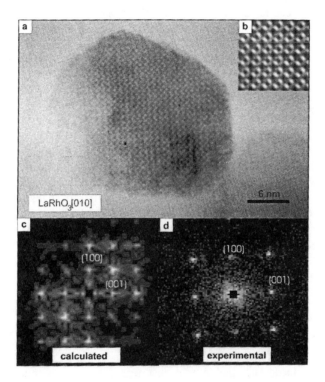

FIGURE 19.10. (a) HRTEM Image of a Particle of $LaRhO_3$ in a 10%Rh/La_2O_3 Catalyst Sample after Heating in He at 927°C. (b) Simulated HRTEM Image for the $LaRhO_3$ Structure. (c) DDP Calculated from this Simulated Image. (d) DDP Calculated from the Experimental Image.

and/or Tb) and a structural model was constructed. The image simulated using this model shows very good agreement with the experimental image, and is shown in Figure 19.7(d). A Beryl-type crystal morphology was employed in order to recreate the facetting of the crystal observed by HRTEM.

Since HRTEM can provide only crystallographic information, in a case where lattice sites in a particular crystal structure may be occupied by more than one similar atom, it is unable to provide details of the ratio of these atoms. The EELS technique was employed[31] to solve this problem. It was found that the ratio of Tb:Ce in the bulk of the mixed oxide support was about 1:4, as expected. However, in the intermetallic crystals, it was much lower. This implies preferential incorporation of Ce into this phase with only a trace of Tb incorporated.

Again, this is a useful illustration of how HRTEM is more powerful when used in combination with other techniques which, in the case of EELS, can be provided built onto the microscope.

19.6.2. Formation of Mixed Oxides

In a further example of a solid state reaction between a supported phase and the support, this time leading to the formation of a spinel mixed metal oxide, a VMgO catalyst containing 14 wt.% V was studied. The catalyst, as prepared, was found to consist of three phases: discrete orthovanadate ($Mg_3V_2O_8$) particles, the MgO support and a disordered V-containing surface layer. Figure 19.8(a) shows a HRTEM image of the VMgO catalyst after exposure to pure propane at 500°C for 4 h.[35,36] The arrangement of the periodic patterns seen in the main image, and shown on an expanded scale in Figure 19.8(b), represent inter-planar spacings of 0.495 and 0.469 nm at an angle of 68° (see DDP in Figure 19.8(c)). These values could not be matched to the known inter-planar spacings and angles of any of the simple oxides of vanadium. It did appear, however, that there was a match with the MgV_2O_4 spinel phase. To confirm this assignment, a structural model was built consisting of a thin layer of spinel supported on the MgO support, as shown in Figure 19.8(d). Images simulated using this model showed striking similarities with the experimental image (Figure 19.8(e)) and DDPs calculated from it gave rise to the expected pattern of spots (Figure 19.8(e)). By reference to the structural model in Figure 19.8(d), it is seen that the spinel phase grew epitaxially on the MgO surface. The mismatch between the spinel and the MgO planes in contact at the interface was calculated to be only 0.2%. Indeed, the structure of the oxygen ion sub-lattice is continuous across the interface between these two phases.

19.6.3. Dissolution and Re-precipitation of the Support Material

Hicks *et al*[37] reported SMSI-like effects on the activity of a Pd/La_2O_3 catalyst after reduction at 300°C. This is a surprising conclusion, since La_2O_3 is widely understood to be essentially unreduced after such a mild treatment. A subsequent study[38] of a 10 wt.% Rh/La_2O_3 catalyst concentrated on elucidating the changes in composition and structure during the catalyst preparation steps and after a range of redox treatments. The authors found that the metal impregnation step caused huge changes in the micro-scale morphology of the support material as observed by SEM. HRTEM was employed to characterise the phases present. It appeared that significant dissolution of the basic La_2O_3 support occurred in the strongly acidic solution containing the Rh precursor and that the Rh metal and La compounds, including oxycarbonates and hydrated species, were subsequently co-precipitated during drying of the catalyst. A typical such polycrystalline particle after undergoing a mild reduction treatment (in H_2 at 200°C for 1 h) is shown in Figure 19.9. The absence of any visible Rh-containing phases supports this co-precipitation explanation since it would be expected that Rh would be very finely dispersed throughout the bulk of these particles. Figure 19.10(a) provides further evidence for such intimate mixing of Rh and La species since it shows an HRTEM image of a particle of the mixed oxide, $LaRhO_3$. This phase was observed in the Rh/La_2O_3 catalyst after a severe thermal treatment (927°C in He). When the Rh/La_2O_3 catalyst was reduced at a higher temperature, 500°C, prior to heat treatment in helium at 900°C, segregation of Rh from the bulk occurred resulting in the formation of Rh particles at the surfaces. Both the size and number of these particles increased with increasing reduction temperature. Furthermore, these particles were decorated with La_2O_3 material from the support as seen in the image and the DDP in Figure 19.11. The authors concluded that the

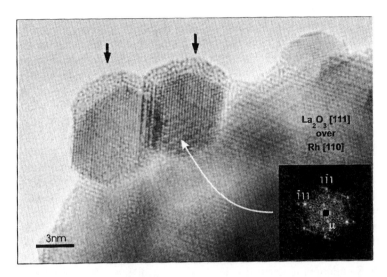

FIGURE 19.11. HRTEM Image of 10%Rh/La$_2$O$_3$ Catalyst Sample after Reduction at 500°C Followed by Heating in He at 900°C Showing Rh Particles Decorated with Support Material (Arrowed). Inset DDP Contains Patterns Attributed to Rh Particle Overlaid with La$_2$O$_3$, and Includes Moiré Spots (Labelled µ).

more severe reduction treatments led to the recrystallisation of La$_2$O$_3$ and that during this process Rh segregated out, forming Rh particles at the surfaces which were decorated with the support material from which they had emerged.

The low activity of La$_2$O$_3$–supported metal catalyst systems after low temperature reduction was therefore assigned to the general inaccessibility of the Rh because of encapsulation of the Rh atoms within the bulk of large polycrystalline particles formed during re-precipitation of the support material in the metal impregnation step. HRTEM observations performed on catalyst samples after various carefully controlled physicochemical treatments was able to uncover this effect.

19.7. SUMMARY

By considering a number of different HRTEM studies of a range of catalyst systems in some depth, we hope to have illustrated the great utility of the technique for elucidating the nano-structure of catalyst systems containing supported phases. Several of the studies addressed clearly demonstrate the incredible sensitivity of HRTEM and its ability to determine the dimensions, morphology, phase and orientation of nano-size supported particles and of supported monolayer structures has been shown.

We have seen that careful measurement of metal particle sizes can provide accurate estimates of metal dispersion. These can be compared with values obtained using chemical techniques to provide insights into the changes in catalyst structure and activity caused by thermo-chemical treatments. It is well known that certain crystallographic planes of catalytic phases are more active for certain reactions than others. It may also be that the Strong Metal-Support Interactions (and metal particle-mediated reductions and oxidations of the support) are also more or less efficient depending on the crystal planes in contact at the interface between the metal particle and the support. Again, HRTEM is invaluable in allowing us to determine the morphology and facetting of active metal and oxide phases, as well as the nature of the interface with the support. This information from the micro- and nano-scale provides great insight into the activity and selectivity of catalytic processes. We have seen that HRTEM is very sensitive to nanoscale modifications in structure brought about by thermo-chemical treatments. It is possible to identify decorating layers on metal nano-particles which are perhaps only one unit cell thick. This resolution of the composition and shape of tiny structures is also possible in supported oxide/oxide systems. Thin layers of one oxide phase lying on another can be identified and the relationship between the two structures at the interface can be determined with incredible precision. The identification of unusual phases, such as the Ce-Pt intermetallic described above, would be very difficult, if not impossible, using 'macroscopic' techniques such as XRD because of the very small size of the crystallites formed and their scarcity in the sample. In our final example, we saw that HRTEM allows us to probe the catalyst preparation process itself. The technique, when coupled with Image Simulation, is unrivalled in its ability to identify the crystal phase of small particles and also to determine the relative dispositions and crystallographic relationships between the phases present in catalyst materials.

Considering the current high interest in the application of HRTEM in catalysis as well as the rate of advancement of the technology, particularly the development of instruments able to operate at atomic resolution with *in situ* environmental cells, this is sure to be a field of increasing importance into the foreseeable future.

REFERENCES

1. P. L. Gai and E. D. Boyles, in: *Electron Microscopy in Heterogeneous Catalysis* (Institute of Physics, 2003).
2. Special issue: *HRTEM for Catalysis, Catal. Today* **23**(3) (1995).
3. Special issue: *Characterisation of Catalysis, Ultramicroscopy* **34**(1&2) (1990).
4. Special issue: *Electron Microscopy and the Characterisation of Heterogeneous Catalysts, Top. Catal.* **21**(4) (2002).
5. P. L. Gai, Developments in *In Situ* Environmental Cell High Resolution Electron Microscopy and Applications to Catalysis, *Top. Catal.* **21**(4), 161-173 (2002).
6. S. Bernal, F. J. Botana, J. J. Calvino, C. López-Cartes, J. A. Pérez-Omil and J. M. Rodríguez-Izquierdo, The Interpretation of HREM Images of Supported Metal Catalysts using Image Simulation: Profile View Images, *Ultramicroscopy* **72**, 135-164 (1998).
7. J. M. Gatica, R. T. Baker, P. Fornasiero, S. Bernal and J. Kašpar, Characterisation of the Metal Phase in NM/Ce$_{0.68}$Zr$_{0.32}$O$_2$ (NM: Pt and Pd) Catalysts by Hydrogen Chemisorption and HRTEM Microscopy: A Comparative Study, *J. Phys. Chem. B* **105**, 1191-1199 (2001).
8. C. Force, A. Ruiz Paniego, J. M. Guil, J. M. Gatica, C. López-Cartes, S. Bernal, J. Sanz, Metal Sintering in Rh/Al$_2$O$_3$ Catalysts Followed by HREM, ^{1}H NMR and H$_2$ Chemisorption, *Langmuir* **17**, 2720-2726 (2001).
9. D. B. Williams and C. B. Carter, in: *Transmission Electron Microscopy, a Textbook for Materials Science* (Plenum Press, New York, 1996), 3-17.
10. D. B. Williams and C. B. Carter, in: *Transmission Electron Microscopy, a Textbook for Materials Science* (Plenum Press, New York, 1996), 177-189.

11. C. Hammond, *Introduction to Crystallography* (Oxford University Press, Royal Microscopical Society, Oxford, 1990).
12. J. J. Hren, J. I. Goldstein and D. C. Joy, *Introduction to Analytical Electron Microscopy* (Plenum Press, New York, 1979).
13. *Electron Microscopy- Principles and Fundamentals* (S. Amelinckx, D. van Dyck, J. van Landuyt, G. van Tendeloo, ed.s, VCH, Weinheim, 1997).
14. J. C. H. Spence, *Experimental High Resolution Electron Microscopy* (Oxford University Press, Oxford, New York).
15. J. F. Watts, *An Introduction to Surface Analysis by Electron Microscopy* (Oxford University Press, Royal Microscopical Society, Oxford, 1990).
16. R. F. Egerton, Application of Electron Energy-Loss Spectroscopy to the Study of Solid Catalysts, *Top. Catal.* **21**(4), 185-190 (2002).
17. D. J. Smith, W. Glaisher, P. Lu and M. R. McCartney, Profile Imaging of Surfaces and Surface-Reactions, *Ultramicroscopy* **29**, 123-134 (1989).
18. P. B. Hirsch, A. Howie, R. B. Nicholson, D. W. Pashley and M. J. Whelan, in: *Electron Microscopy of Thin Crystals* (Butterworths, Sevenoaks, U.K.).
19. L. D. Marks and D. J. Smith, Direct Surface Imaging in Small Metal Particles, *Nature* **303**, 316-317 (1983).
20. D. J. Smith, The Realisation of Atomic Resolution with the Electron Microscope, *Rep. Prog. Phys.* **60**, 1513-1580 (1997).
21. J. M. Gatica, R. T. Baker, P. Fornasiero, S. Bernal, G. Blanco, and J. Kašpar, Rhodium Dispersion in a $Rh/Ce_{0.68}Zr_{0.32}O_2$ Catalyst Investigated by HRTEM and H_2 Chemisorption, *J. Phys. Chem. B* **104**(19), 4667-4672 (2000).
22. S. Bernal, J. J. Calvino, M. A. Cauqui, J. A. Pérez-Omil, J. M. Pintado and J. M. Rodríguez-Izquierdo, Image Simulation and Experimental HREM Study of the Metal Dispersion in Rh/CeO_2 Catalysts. Influence of the Reduction/Reoxidation Conditions, *Appl. Catal. B* **16**, 127-238 (1998).
23. A. Borodzinski and M. Boranowska, Relation Between Crystallite Size and Dispersion on Supported Metal Catalysts, *Langmuir* **13**, 5613-5620 (1997).
24. S. Bernal, R. T. Baker, A. Burrows, J. J. Calvino, C. J. Kiely, C. Lópex-Cartes, J. A. Pérez-Omil and J. M. Rodríguez-Izquierdo, Structure of Highly Dispersed Metals and Oxides: Exploring the Capabilities of High-Resolution Electron Microscopy, *Surf. Interface Anal.* **29**, 411-421 (2000).
25. S. Bernal, J. J. Calvino , M. A. Cauqui , J. M. Gatica , C. Larese , J. A. Pérez-Omil and J. M. Pintado, Some Recent Results on Metal/Support Interaction Effects in NM/CeO_2 (NM : Noble Metal) Catalysts, *Catal. Today* **50**, 175-206 (1999).
26. S. Bernal, F. J. Botana, J. J. Calvino, G. A. Cifredo, J. A. Pérez-Omil and J. M. Pintado, HREM Study of the Behaviour of a Rh/CeO_2 Catalyst under High-Temperature Reducing and Oxidising Conditions, *Catal. Today* **23**, 219-250 (1995).
27. S. Bernal, J. J. Calvino, M. A. Cauqui, J. M. Gatica, C. López-Cartes and J. M. Pintado, Chemical and Nanostructural Aspects of the Preparation and Characterisation of Ceria and Ceria-Based Mixed Oxide Supported Catalysts, in: *Catalysis by Ceria and Related Materials* edited by A. Trovarelli (Imperial College Press), Chapter 4, 85-168 (2002).
28. S. J. Tauster, S. C. Fung and R. L. Garten, Strong Metal-Support Interactions. Group 8 Noble Metals Supported on Titanium Dioxide, *J. Am. Chem. Soc.* **100**, 170-175 (1978).
29. S. J. Tauster, Strong Metal-Support Interactions, *Accounts Chem. Res.* **20**(11), 389-394 (1987).
30. G. L. Haller and D. E. Resasco, Metal Support Interaction – Group VIII Metals and Reducible Oxides, *Adv. Catal.* **36**, 173-235 (1989).
31. S. Bernal, J. J. Calvino, M. A. Cauqui, J. M. Gatica, C. López-Cartes, J. A. Pérez-Omil and J. M. Pintado, Some Contributions of Electron Microscopy to the Characterisation of the Strong Metal-Support Interaction Effect, *Catal. Today* (in press).
32. S. Bernal, J. J. Calvino, J. M. Gatica, C. Larese, C. López-Cartes and J. A. Pérez-Omil., Nanostructural Evolution of a Pt/CeO_2 Catalyst Reduced at Increasing Temperatures (473-1223 K): A HREM Study, *J. Catal.* **169**, 510-515 (1997).
33. S. Bernal, J. J. Calvino, M. A. Cauqui, J. M. Gatica, C. Larese, J. A. Pérez-Omil and J. M. Pintado, Some Recent Results on Metal/Support Interaction Effects in NM/CeO_2 (NM : Noble Metal) Catalysts, *Catal. Today* **50**, 175-206 (1999).

34. G. Blanco, J. J. Calvino, M. A. Cauqui, P. Corchado, C. Larese, C. López-Cartes and J. A. Pérez-Omil, Nanostructural Evolution under Reducing Conditions of a Pt/CeTbO$_x$ Catalyst: A New Alternative System as a TWC Component, *Chem. Mater.* **11**, 3610-3619 (1999).

35. A. Burrows, C. J. Kiely, J. A. Pérez-Omil, J. J. Calvino and R. W. Joyner, Structural Characterisation of the VMgO Catalyst System, *Inst. Phys. Conf. Ser.* **153**, 395-398 (1997).

36. A. Burrows, C. J. Kiely, J. Perregard, P. E. Hojlund-Nielsen, G. Vorbeck, J. J. Calvino and C. López-Cartes, Structural Characterisation of a VMgO Catalyst Used in the Oxidative Dehydrogenation of Propane, *Catal. Lett.* **57**(3), 121-128 (1999).

37. R. F. Hicks, Q. -J. Yen, A. T. Bell, Effects of Metal Support Interactions on the Chemisorption of H$_2$ and CO on Pd/SiO$_2$ and Pd/La$_2$O$_3$, *J. Catal.* **89**, 498-510 (1984).

38. S. Bernal, J. J. Calvino, C. López-Cartes, J. A. Pérez-Omil, J. M. Pintado, J. M. Rodríguez-Izquierdo, K. Hayek and G. Rupprechter, Nanostructural Evolution of High Loading Rh/Lanthana Catalysts through the Preparation and Reduction Steps, *Catal. Today* **52**, 29-43 (1999).

20

Solid State Physics and Synchrotron Radiation Techniques to Understand Heterogeneous Catalysis

D. C. Bazin[*]

20.1. INTRODUCTION

Heterogeneous catalyst[1,2] consists of a large specific area oxide ($200m^2$/g typically) at the surface of which nanometer-scale clusters[3,4] are anchored. The understanding of the adsorption process of simple molecules at the surface of these particular aggregates (containing a small number of atoms) constitutes the core of this paper.[5,6] The ultimate goal is to correlate their structure at the atomic level with catalytic properties[7] in order to simulate (understand/predict) their activity/selectivity.[8,9] No attempt is made to review the recent experimental data, since excellent survey articles of this nature already exist.[10,11]

Some of the major breakthroughs performed in heterogeneous catalysis are based on research using synchrotron radiation related techniques.[12,13] Among them, X-ray Absorption Spectroscopy (XAS)[14] has been shown to be a very powerful tool for exploring the structure and electronic properties heretofore inaccessible to conventional techniques. Anomalous Wide Angle X-ray Scattering (AWAXS)[15] is also an elegant technique especially effective for probing the range order beyond the first coordination sphere of a metal selected by its edge energy. Attention will be paid to the complementarity[16] between these techniques in order to define a methodology in the building of a structural model.[17]

[*] D. C. Bazin, LURE, Bât 209D, Université Paris XI, B. P. 34, 91898 Orsay Cedex France

To illustrate our purpose, the selected examples treated here are related to reforming,[18] to the postcombustion[19] and to the Fischer-Tropsch processes.[20] More precisely, in the case of reforming catalysts, the experimental results[21] obtained on the sulfuration process of the PtRe bimetallic system give direct structural evidence of the role of the second metal i.e. the rhenium. More precisely, in the case of the Pt/Al$_2$O$_3$ monometallic system, the sulphur modifies completely the structure of the platinum clusters. If we consider now the PtRe bimetallic system, the sulphur provokes no major change in the metal cluster structural parameters.

Regarding the post combustion process, a combined approach[22] based on XAS and AWAXS leads to a fine description of the cation distribution, the electronic state of the metal atoms, and the size of metal oxide clusters for a highly dispersed spinel phase, here the supported system ZnAl$_2$O$_4$/Al$_2$O$_3$. The complete set of results shows a significant lack of occupancy on the tetrahedral zinc site. Also in situ studies reveal a significant increase of the cell parameter as well as a dramatic increase of the Debye-Waller factor associated with the vibration of the zinc-zinc pairs for a spinel-like environment of zinc when such a phase is submitted to gas-solid reactions such as those employed for automotive exhaust control. An explanation of these two effects is given on the basis of the incorporation of a light atom (such as oxygen atoms) coming from the reactive gases into the solid via the defective properties of the ZnAl$_2$O$_4$/Al$_2$O$_3$ spinel surface.

In the last example dedicated to the Fischer-Tropsch reactions, one of the original points was to combine in situ H$_2$ reduction (under pure H$_2$ between 300°C and 650°C) of calcined silica-supported Co catalysts with soft X-ray absorption spectroscopy at the Co L$_{II,III}$ edge.[23] We will see that the experimental data can be utilised to link structural characteristics to the catalytic activity.

Finally, we would like to introduce the possibility of predicting the activity of a catalyst[24] through the latest breakthroughs in the calculation of the electronic structure of nanometer-scale metallic entities.[25] The starting point is given by the experimental results obtained on postcombustion catalysts.[26] Following a fundamental approach, we have considered the influence of 1% NO/balance N$_2$ over Pt clusters deposited on γ-Al$_2$O$_3$. It seems that the introduction of NO leads to the sintering of particles. The question of adsorption of simple molecules such as NO on nanometer-scale metallic aggregates is then discussed in terms of energy consideration.[27] More precisely, we suggest that a relationship exists between the adsorption mode of NO (*dissociative or molecular*) and the behaviour of the metal nanoparticles (*sintering or disruption*).

20.2. THE INVESTIGATION TOOLS

Hitherto, the usual characterisation techniques, such as transmission electron microscopy, energy-dispersive X-ray emission, X-ray photoelectron spectroscopy and CO chemisorption analysis help to delimit the scientific case but do not precisely show the intimate structure of the metallic particles while the reaction occurs. The emergence of synchrotron radiation centre in the past decade has given a tremendous development of various *in situ* structural and electronic characterisation techniques.[28] Each of them has advantages and limitations related to the specific properties of the materials investigated. In fact, it is their complementary which allows the building of a concise structural model.[16]

Regarding XAS,[29] analysis of the fine structures present after X-ray absorption edges enables a fine description of the very first coordination sphere around an element selected through its absorption edge.[30]

More precisely, the known interatomic distances can be used as fingerprints of the network (face-centred cubic or body-centred cubic). Information about the size of the cluster can also be extracted from the average coordination number since it is well known that the average coordination varies as the size of the cluster increases (Fig. 20.1). Unfortunately, the uncertainty in measuring the coordination number (around 20%) gives a significant error in determining the size.

When looking at the higher shells which are the key for determining the morphology, the signal associated with these shells is very small. For the smallest cuboctahedra (13 atoms), we consider the average number associated with the fourth shell is 0.9 and is similar to the error commonly given to the coordination numbers (+1 in the case of such materials). Thus, for such small clusters, information about morphology can not easily be determined.[31] Note that XAS faces a major difficulty. This technique is insensitive to polydispersity.[32]

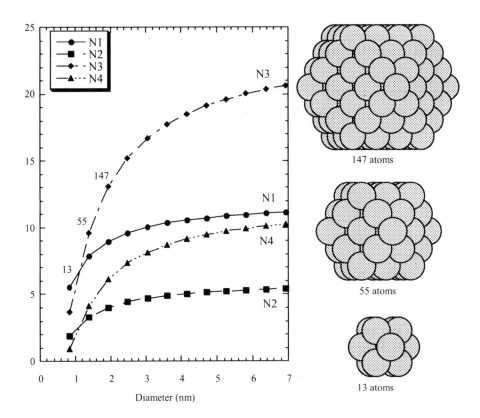

FIGURE 20.1. Variation of the first (N1) through fourth shell (N4) coordination numbers versus the number of atoms in the clusters. Cuboctahedra clusters containing 13, 55 or 147 atoms have been represented.

This technique can also be used to observe the interaction between either the metal and the support or the metal and reactive gases while the reaction occurs.[33] More precisely, this tool offers an unique opportunity to follow the local order around the metal during each step of the preparation procedure namely the impregnation-drying,[34] the calcination[35] and the reduction.[36] Such description at the atomic level leads to a better understanding of the intimate relationship between the structural characteristics of the material and its catalytic activity/selectivity.[37]

Particular attention has to be paid to the X-ray Absorption Near Edge Structure (XANES). First, we consider the analysis of the K edge (photoabsorption process leading to the ejection of 1s electron) of 3d or 4d transition metals. From a practical point of view, numerical simulation of the XANES part of the K absorption spectrum of most elements can be performed through full multiple scattering calculations.[38] Then, on the basis of a linear combination of the XANES spectra of reference compounds,[39] the presence of the different phases present inside the materials can be quantified.

Recently, we have shown that for nanometer-scale metallic clusters, it is not sufficient to consider only the electronic state of the metal of interest to perform a linear combination analysis.[40] In the case of these peculiar materials, special attention has to be paid to different structural parameters, for example the size and morphology of the cluster, the interatomic distance (taking into account contraction/dilatation processes), and the presence of heterometallic bonds (in the case of bimetallic clusters).

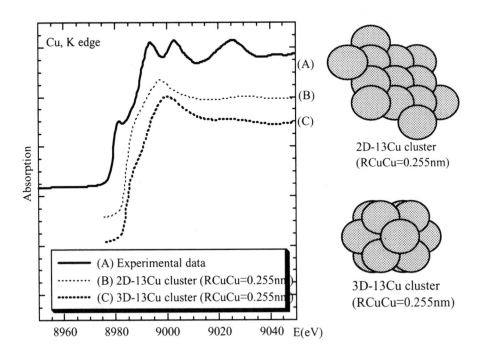

FIGURE 20.2. Near K edge part of the absorption spectrum as measured for the copper foil (A) and as calculated for two clusters of 13 atoms of copper having a 2D (B) or 3D dimension (C).

Numerical calculations are plotted on Fig. 20.2 corresponding to two clusters of 13 atoms with a face-centred cubic network but with either a plane (2D) or a sphere (3D) like morphology. It is clear that the morphology of the cluster constitutes a significant parameter, leading to major modification of the shape and the position of the edge. Thus, it is definitely not sufficient to consider only the electronic state of the metal of interest i.e. it is not possible to simulate the XANES part of the absorption spectra associated to a nanometer-scale metallic cluster with the one associated to the corresponding metallic reference. In fact, the quantitative measurement of the structural parameters coming from XAS analysis constitutes an invaluable starting point for the linear combination of the XANES spectra. The fact that major results coming from the emergence of dynamical studies[41] have now been obtained will lead to significant breakthroughs in the understanding of the genesis of nanometer-scale entities.[42]

Regarding the $L_{II,III}$ edges of 5d metals (photoabsorption process leading to the ejection of 2p1/2 for L_{II} or 2p3/2 electron for L_{III}), we have shown in the case of nanometer-scale platinum clusters that a strong correlation (Fig. 20.3) exists between the intensity of the large threshold spikes which appear at the $L_{II,III}$ edges of transition metals and the size of the cluster.[43] Thus, at least two physical phenomenon can affect the intensity of the large threshold spikes: the size of the of the cluster which can be considered as an intrinsic effect (density of state of nanometer-scale platinum cluster is far from the bulk one) and a possible charge transfer between the cluster and the support which can be considered as an extrinsic one.

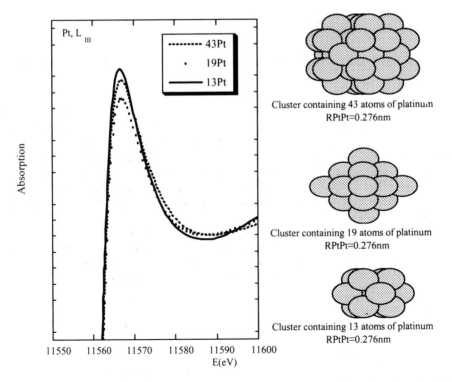

FIGURE 20.3. Intensity of the large threshold spikes which appear at the platinum L_{III} edges for different clusters containing 13, 19 or 43 atoms.

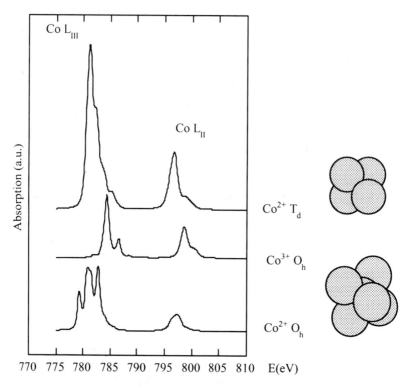

FIGURE 20.4. Set of numerical simulations of the Co L edge absorption spectra using the multiplet theory corresponding to the configurations of Co^{2+} T_d, Co^{3+} O_h, and Co^{2+} O_h.

In the case of 3d and 4d transition metals, while a shoulder is generally measured at the K-edge, fine structures are present at the L edges.[44] The recent advances in theoretical background related to soft X-ray experiments have motivated a set of numerical simulations (Fig. 20.4) corresponding to the following configurations for cobalt atoms : Co^{2+} T_d, Co^{3+} O_h, and Co^{2+} O_h. These calculations show the great sensitivity of 2p spectroscopy to the valence state of the cobalt atoms as well as to the symmetry of the sites. Note that the exact position of these spectra versus photon energy is not determined from theory.

Moreover, X-ray absorption spectra can be collected at the K edge of light elements such carbon, nitrogen or oxygen. Thus, the adsorption process of molecules of major interest in heterogeneous catalysis can be studied completely from the two actors of the chemical bond leading to a precise knowledge of intermediate species.

Regarding AWAXS, some basic elements of X-ray diffraction have been already presented.[45] Here, we will just introduced some opportunities given by WAXS (Wide Angle X-ray Scattering). In fact, it is easy to calculate the diffraction diagram (Fig. 20.5) of the metallic part of a catalyst composed of a collection of nanometer-scale metallic clusters with known geometry but random orientation and position. The splitting between the diffraction peaks is extremely sensitive to the size of the clusters (here cuboctahedra) allowing a direct measure of the cluster size. For example, the separation between the 111 and the 200 diffraction peaks can be used as a determination of the size

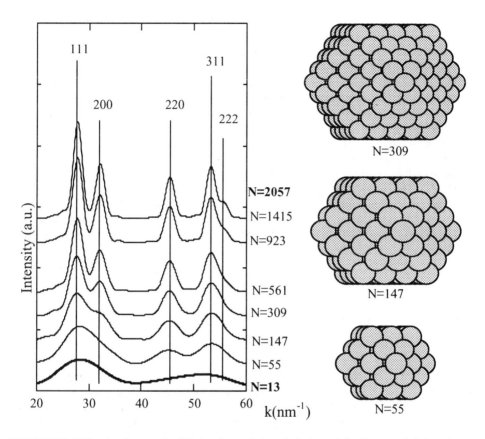

FIGURE 20.5. Diffraction diagrams for different clusters (cuboctahedra) containing N atoms of platinum (some of them have been represented).

if the number of atoms inside the cluster is under 1000 (then we can consider the splitting between the 311 and 222 diffraction peaks). Note that such numerical simulations lead also to significant information regarding the morphology of the particle.[46]

20.3. EXPERIMENTAL DEVICES DEDICATED TO HETEROGENEOUS CATALYSIS

Regarding the reforming process, we have developed a device where the sample is placed in a oven capable of withstanding hydrogen pressure (>20 atm.) and temperature (<500°C).[47] Moreover, it is possible to inject normal heptane over the catalyst (while the total H_2 pressure is equal to 5 atm.). Then, a complete reaction cell that simulates DeNO$_x$ experimental conditions has been developed.[48]

In this device, the sample is placed at high temperature under the flow of a complex mixture of reactive gases (NO, C_3H_6, O_2, N_2). Precise control of the gas flow is maintained by means of mass flow controllers, and the safety of the set-up is assured by

CO, CO_2 and NO detectors directly relayed to an electronic device, which is itself linked to the various mass flow controllers.

One of the specific aspects of postcombustion is a high flux for the reactive gas. Such a characteristic is not really compatible with the geometry of a classical furnace. In order to perform the XAS experiments in real conditions, we must define a complete new sample holder.[49] Basically, this sample holder is a pipe made of pyrolitic boron nitride, the reactive gas coming from one extremity, the other one being the gas exit. Thus, a high flux of the reactive gases can go through the sample as is the case for industrial converters. Moreover, due to the fluorescence mode, a particular geometry is given to the tube where the monochromatic beam arrives on the sample. At this point, the thickness of the pipe is close to 0.15 mm, leading to a transmittance equal to 0.924 at 10000. eV. This low value has required the use of high technology to build the sample holder for which high purity as well as chemical stability are required. In fact, the sample holder is produced by chemical vapour deposition onto machined graphite mandrels at a temperature between 1800 and 1900°C. The pyrolitic boron nitride forms upon the graphite mandrel as a coating and is allowed to grow up to a typical thickness of 0.8 mm. On cooling, the coating is separated from the mandrel to produce the sample holder.

20.4. SELECTED EXAMPLES OF CATALYTIC STUDIES

20.4.1. Reforming

At the outset, in collaboration with J. Lynch (Institut Français du Pétrole, Rueil-Malmaison, France) interesting results have been obtained on the PtRe and PtRh[50] bimetallic systems. The originality of this study was based on the following of the first coordination sphere of each metal during the first stages of preparation procedure namely the impregnation-drying, the calcination and the reduction. Regarding the reduction step, an epitaxy phenomenon was observed[51] through an enlargement of the metal-metal distance whose value approaches the (111) interplanar distance of alumina (2.85 Å).

Then, significant breakthroughs have been achieved by taking into account a major experimental parameter in heterogeneous catalysis namely the pressure.[52] In fact, XAS has been used to follow in situ the structural evolution of a chlorinated and non-chlorinated Pt/Al_2O_3 catalyst during reduction in the temperature range of 300-500°C. The experimental data shows that smaller metal clusters are formed from the hydrogen reduction of the chlorinated catalyst, in contrast to the larger cluster formed from the non-chlorinated one.

At 460°C, the total hydrogen pressure was raised to 5 atm. and n-heptane was injected over the samples. XAS measurements at the Pt edge were carried out while hydrocarbon conversion was monitored with a gas chromatograph. We observe the rapid formation of a carbon-platinum bond. Note that this is unmodified while turnover rates and selectivities indicate evidence for deactivation.

From this structural information supplied by XAS, correlated with the data obtained from gas chromatography, we find that our results are consistent with a model proposed by others where deactivation is due to the build-up of a multilayer of carbon.[53]

The sulfuration process of a reforming catalyst has been also studied. More precisely, the research was focused on Pt and $PtRe/Al_2O_3$ systems.[21,54] For the monometallic system (Fig. 20.6), the original sample presents an average Pt-O coordination number of

6.2. During the H_2 treatment, the platinum atom is reduced to the metallic state, with a Pt-Pt coordination number of 4.5 (R_{PtPt}= 2.72Å). For the Pt-O coordination, we measure a number of 0.4 which is negligible considering the precision of ± 0.5 for coordination numbers measured by XAS showing thus that the Pt particles are fully reduced.

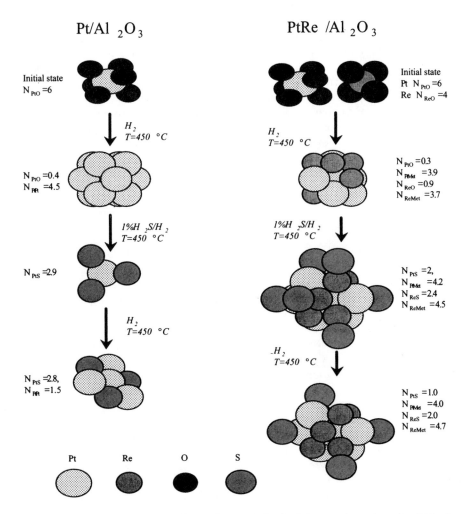

FIGURE 20.6. Schematic representation at the atomic level of the experimental results obtained on the sulfuration of the Pt/Al$_2$O$_3$ and the PtRe/Al$_2$O$_3$ catalytic systems.

After sulfuration in flowing 1%H$_2$S in H$_2$ at 723K, the platinum neighbouring environment is essentially formed by sulphur atoms. The metallic particles disappear. On average, we have found a coordination number for Pt-S of 2.9 at a distance of 2.32Å. Finally, after a second reduction in flowing H$_2$ at 723K, the environment of platinum atoms is composed of 2.8 sulphur atoms at 2.30Å and 1.5 platinum atoms at 2.71Å.

The same set of experiments has been made on the bimetallic PtRe system (Fig. 20.6). At the beginning, the environment of the two metals is composed essentially of oxygen, six for the platinum and four for the rhenium. The reduction process introduces a metal (either platinum or rhenium) in the local order around the two metals.

Metallic clusters are thus generated, containing very few atoms as can be seen from the low metal coordination numbers. Contact with H_2S does not change the structural characteristics of the metallic clusters, the metal coordination numbers remaining the same. Sulphur atoms are located around both metals. If the catalyst is then submitted to pure hydrogen, part of the sulphur in the vicinity of the platinum is removed, the rhenium keeping its sulphur environment.

If we compare the results obtained on the Pt/Al_2O_3 monometallic system to the ones obtained on the $PtRe/Al_2O_3$ bimetallic system, the fact that in the bimetallic system, platinum-metal bonds still exist after the sulfidation underlines the role played by the rhenium atoms. The presence of the second metal stabilizes the metallic cluster.

In addition, the results show clearly that the concept of selective poisoning of rhenium sites is correct, the sulfur bond to rhenium shows a much greater stability than that to platinum.

20.4.2. Post Combustion

The study of automobile catalytic converters has recently received a burst of attention arising from the importance of such experimental device from an ecological point of view. The decomposition of nitrogen oxides can be observed on ternary AB_2O_4 spinel oxides. In collaboration with A. Seigneurin (Rhodia, Aubervilliers, France), a combined approach based on XAS and AWAXS[22,55] leads to a fine description of the cation distribution, the electronic state of the metal atoms, and the size of metal oxide clusters for a highly dispersed spinel phase, here the supported system $ZnAl_2O_4/Al_2O_3$. Through numerical simulations of the XAS data and calculations of the differential diffraction intensities, a model of the zinc distribution inside the alumina particles is proposed, in which the zinc atoms are mostly located at the surface of the alumina particle with a concentration gradient between the surface and the particle center.

In situ studies reveal a significant increase of the cell parameter as well as a dramatic increase of the Debye-Waller factor associated with the vibrations of the zinc-zinc pairs for a spinel-like environment of zinc when such a phase is submitted to gas-solid reactions such as those employed for automotive exhaust control. An explanation of these two effects is given on the basis of the incorporation of a light atom (such as O) coming from the reactive gases into the solid via the defective properties of the $ZnAl_2O_4/Al_2O_3$ spinel surface.

In fact, the different results we have obtained on the two systems $ZnAl_2O_4/Al_2O_3$ and $SnO_2\text{-}ZnAl_2O_4/Al_2O_3$ have shown the necessity to collect the data during the chemical process, the physico-chemical state of the catalyst being the same before and after the reaction. Taking into account the complete set of results, we can assume thus that only a dramatic lack of occupancy on the metal site favours an incursion of light atoms in the network. Such behaviour can explain in return the expansion of the crystallographic cell as well as a significant increased of the Debye-Waller factor associated to zinc-zinc pairs.

Note that the NO adsorption on Zn/Al_2O_3 powder has been also investigated by NEXAFS study at nitrogen K edge.[56]

20.4.3. The Fischer-Tropsch Process

Conducted under the direction of Prof. L. Guczi, this research dedicated to the Fischer-Tropsch is part of an international program named "Surface structure and reactivity relationship in CO/H_2 reaction over cobalt based catalyst". More precisely, structural/electronic investigations have been conducted on several catalytic systems namely Co/SiO_2,[57] $CoPd/SiO_2$,[58] $Ru-Co/NaY$,[59] $PtCo/NaY$ or $/Al_2O_3$,[60] and finally $ReCo/NaY$ or $/Al_2O_3$.[61]

Regarding the monometallic system,[23] the original goal was to combine in situ H_2 reduction (under pure H_2 between 300 and 650°C) of calcined silica-supported Co catalysts with soft X-ray absorption spectroscopy at the Co $L_{II,III}$ edge. Using reference compounds as well as numerical simulations based on the multiplet theory (Fig. 20.4), the structural transition occurring on Co_3O_4 clusters to give CoO species was established (Fig. 20.7).

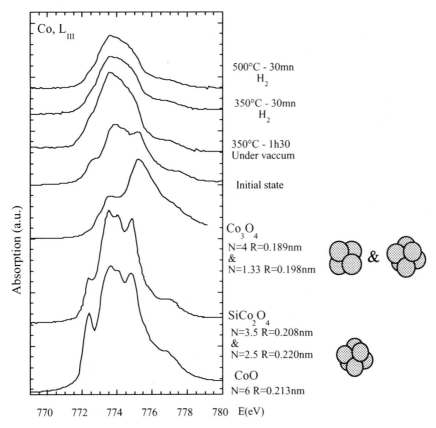

FIGURE 20.7. Evolution of the Cobalt L_{III} edge during the reduction under hydrogen done at different temperatures for the catalyst a-Co/SiO_2.

Note that disappearance of the different features associated with Co^{2+} O_h symmetry was observed, probably due to disorder in the first coordination sphere of the cobalt atoms. Here, we arrive at the limit of the theoretical formalism currently used in the analysis of transition metal L-edge absorption. Thus, the significant information gained from this particular result can be utilised to link structural characteristics to the catalytic activity which is also a challenge for theoretical physics.

20.5. USING SOLID STATE PHYSICS TO UNDERSTAND HETEROGENEOUS CATALYSIS

The starting point of this approach is given by a study of the reaction between carbon oxide and nitrogen oxides (NO$_x$) over Pt-containing catalyst particle in order to obtain CO$_2$ and NO$_2$.[62] In order to simplify the catalytic process,[63] we have considered the interaction of different reactant gases over Pt clusters deposited on γ-Al$_2$O$_3$.[64] The influence of 1% NO/balance N$_2$ was studied as a function of the temperature on a 1 wt.% Pt/γ-Al$_2$O$_3$. When the catalyst is pre-reduced before the introduction of NO, the sintering of particles is considerable above 200°C.[65] This behaviour seems to be independent of the support, the coalescence process being observed in the case of zeolite.[66]

At this step of the analysis, an interesting point of comparison is given by the NO adsorption on metallic surfaces. It is well known that the ability of transition metal surfaces to dissociate simple molecules (N$_2$, O$_2$, NO) decreases across the metal transition series.[25] Recently, Brown[67,68] suggested a correlation between the propensity for dissociation of the NO monomer at low coverage and the melting points of the transition metals.

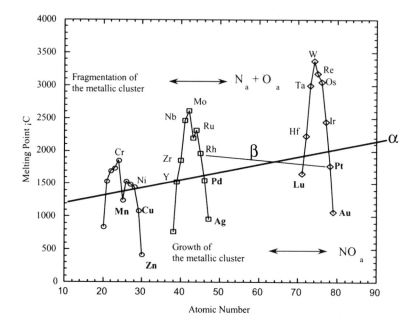

FIGURE 20.8. Relationship between the adsorption mode of NO (dissociative or molecular) and the behaviour of the metal nanoparticles (sintering or disruption). A line α separates the two adsorption modes for NO. A line β is defined for the PtRh bimetallic catalyst.

Taking account of the latest breakthrough performed in the calculation of the electronic structure of nanometer-scale metallic clusters (see reference 69 for more details) as well as the suggestion of Brown for metallic surfaces, we have proposed a relationship between the adsorption mode of NO (dissociative or molecular) and the behaviour of the nanometer-scale metallic particles (sintering or disruption). More precisely, a straight line α separates these two possibilities (see Fig. 20.8).

Recently, a publication[70] has been dedicated to the study of interaction of NO with supported Rh nanoparticles. According to XAS[71] in the initial state the immediate neighbourhood of Rh atoms is composed of 8 Rh atoms, with Rh-Rh bonds of distances distance of 2.68Å. After exposure to 4 % NO/He at 313K for 5 seconds, the number of Rh-Rh bonds significantly decreased (from 8 to 2). Oxygen atoms (N_{RhO}=2, R_{RhO}=2.05Å) are present also in the first coordination sphere. This structural description is completely in line with the relationship we have proposed between the mode of adsorption (*dissociative vs. molecular*) and the behaviour of the metal nanoparticles (*sintering or disruption*). Let's try now to consider different questions.

- Is it possible to use this simple model to predict the behaviour of the metal nanoparticles when a mixture of NO+O_2 is considered? Regarding rhodium on which NO molecule underwent dissociative adsorption with the formation of metal-oxygen bond formation, it is possible that the presence of oxygen will not change significantly this simple scheme. In contrast, for platinum the situation is more complex because the temperature plays a significant role. In a recent paper,[72] we have studied the effect of gas mixtures (NO+O_2) with excess oxygen. XAS gives a direct structural evidence of a platinum/oxygen substitution in the platinum environment. However, some Pt-Pt bonds persist till 300°C, which was not the case when oxygen was present alone. Thus, the presence of NO allowed the Pt particles to conserve a metallic character. Note that the relative concentrations of the two gases, as well as the temperature, probably constitute two significant parameters.

We would like to consider now bimetallic catalysts. XAS experiments performed at the Pt L_{III} edge have been carried out on a bimetallic 1wt% Pt-1wt% Rh. For Pt-Rh particles, the sintering process was not observed.[65] It is thus clear that the presence of Rh in the environment of Pt modifies significantly the behaviour of the particles versus the NO adsorption process. In a first approach, we consider a straight line (line β on Fig. 20.8) between platinum and rhodium on the diagram we have already proposed for monometallic system. In the Pt rich region, the behaviour of PtRh bimetallic cluster will be similar to the Pt monometallic system. However, in the Rh rich region, the behaviour of PtRh bimetallic cluster will be similar to the Rh monometallic system. If we consider now the experimental result obtained on the 1 % Pt-1 % Rh bimetallic system, we see that this result is *in line* with the conclusion obtained with this simple approach.

- Is it possible to use this simple model to choose the two metals A and B which constitute the metallic part of the catalyst? If we assume that the catalytic activity depends of the metallic part of the catalyst, the structural characteristics of the AB bimetallic particles must be stable during the adsorption of NO molecules.

Such bimetallic catalyst has to be positioned on the straight line α (i.e. an intersection exists between the line α defined on Fig. 20.8 and a line β defined by the two metals A and B) i.e. one metal has to be positioned above the straight line α while the other one has to be below.

This configuration is possible for the PtRh bimetallic system for example. At the opposite, following this simple relationship, we can assume that the activity of RhRu bimetallic catalyst will change dramatically with time. The NO adsorption will modify significantly the structural characteristics of the bimetallic particles (probably an incursion of oxygen in the vicinity of the two metals).

In summary, although it is clear the important question dealing with the role of different parameters such the temperature, the pressure or the coverage remains unanswered, some insight has been gained through the insertion of physical concepts in the understanding of the catalytic processes.

20.6. SOME PERSPECTIVES AND CONCLUSION

We have just discussed previously activity. What about selectivity? Chirally modified metals are the most effective heterogeneous catalysts for enantioselective hydrogenation reactions.[73] In fact, there is a simple question. Is it possible to design nanometer-scale clusters which correspond to two enantiomers? On Fig. 20.9, we build a monometallic nanotube on which we put some atoms of a second metal.

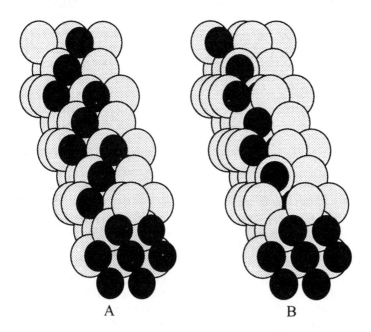

A   B

FIGURE 20.9. Atomic representation of the two clusters A and B which can be viewed as two enantiomers.

From what has been said previously, it is evident that the combination of synchrotron radiation techniques along with the development of theoretical simulations has made possible the efficient detailed study of many chemical reactions. Application of these techniques to the study of Fischer-Tropsch process or post combustion catalytic converters has been particularly fruitful.

Regarding reforming, we have studied two catalytic systems, the monometallic Pt/Al_2O_3 and the bimetallic $PtRe/Al_2O_3$ systems. The complete set of data shows clearly that the concept of selective poisoning of rhenium sites is correct, the sulfur bond to rhenium shows a much greater stability than that to platinum. Also, the presence of rhenium stabilizes the metallic cluster.

The different results we have obtained on the two systems $ZnAl_2O_4/Al_2O_3$ and SnO_2-$ZnAl_2O_4/Al_2O_3$ dedicated to postcombustion have shown the necessity to collect the data during the chemical process, the physico-chemical state of the catalyst. One of the key results is linked to the fact that only a dramatic lack of occupancy on the metal site favours an incursion of light atoms in the network.

Finally, in the case of the Fischer-Tropsch process, we use the recent advances in theoretical background related to soft X-ray experiments to show the great sensitivity of 2p spectroscopy to the valence state of the cobalt atoms as well as to the symmetry of the sites. Also, disappearance of the different features associated with Co^{2+} O_h symmetry was observed, probably due to disorder in the first coordination sphere of the cobalt atoms.

During recent years, we have been witness to significant scientific and technological efforts to produce uniform shape and size nanometer-scale metallic clusters. For example, it has been shown that well-defined platinum particles encapsulated in mesoporous silica can be prepared now.[74] Regarding characterisation techniques, new ones have emerged such as sum frequency generation (SFG) which can be utilized to monitor the substrate and adsorbate structure.[75] Continued developments of the preparation procedure as well as new efficient tools for the investigation of chemical reaction dynamics will provide significant breakthroughs.

ACKNOWLEDGEMENTS

I am indebted to Prof. D. Sayers from North Carolina State University (Raleigh, United-States) and Prof. L. Guczi from the Institute of Isotope and Surface Chemistry (Budapest, Hungary).

The author thanks past and present co-workers, whose names appear in the reference list, for their valuable contributions. I am particularly grateful to Dr. H. Dexpert from the Centre d'Elaboration de Matériaux et d'Etudes Structurales (Toulouse, France), Dr. F. Garin, Dr. G. Maire from the Laboratoire des Matériaux, Surfaces et Procédés pour la Catalyse (Strasbourg, France), Prof. J. Barbier, Prof. C. Kappenstein, Prof. C. Micheaud, Prof. M. Guerin from the Laboratoire de Catalyse en Chimie Organique (Poitiers, France) Dr. J. Lynch, Dr. C. Pichon, Dr. Rebours, Dr. Revel from the Institut Français du Pétrole (Rueil-malmaison, France), Dr. A. Seigneurin, Dr. A. Pourpoint from Rhodia (Aubervilliers, France), Dr. M. Capelle, Dr. F. Maire, Dr. G. Meunier, Dr. R. Noirot from Peugeot-Citroën (Belchamp, France), Dr. J. Moonen from Dutch State Mines (Geleen, Netherlands), Dr. I. Kovacs, Dr. D. Horwath from the Institute of Isotope and Surface Chemistry (Budapest, Hungary) for invaluable discussions on heterogeneous catalysis.

Special thanks are due to Prof. J. J. Rehr from the University of Washington (Seattle, United-States), Dr. D. Spanjaard from the Laboratoire de Physique des Solides (Paris XI University, France), Dr. M. C. Desjonqueres from Centre d'Etudes Atomiques (Saclay, France), Dr. G. Tréglia, Dr. C. Mottet from the Centre de recherche sur les mécanismes de croissance crystalline (Marseille, France), to Dr. E. Elkaim, Dr. A.M. Flank, Dr. M. Gailhanou, Dr. C. Lafon, Dr. P. Lagarde, Dr. J. P. Lauriat, Dr. Ph. Parent, Dr. S.

Rouziere, Dr. D. Thiaudière, from LURE (Laboratoire pour l'Utilisation du Rayonnement Electromagnétique, Orsay, France) for invaluable discussions.

Finally, grateful acknowledgment is also made to chemical department of the "Centre National de la Recherche Scientifique" for financial support of the different programs of which this work forms a part.

REFERENCES

1. G. A. Somorjai and K. McCrea, Roadmap for catalysis science in the 21st century: a personal view of building the future on past and present accomplishments, *Applied Cat. A: General* **222**, 3-18 (2001).
2. P. W. Jacobs and G. A. Somorjai, Conversion of heterogeneous catalysis from art to science: the surface science of heterogeneous catalysis, *J. of Molecular Cat. A : Chemical* **131**, 5-18 (1998).
3. J. Friedel, Physics of Metals 1, Ed. Cambridge University Press, 1978.
4. J. H. Sinfelt, Bimetallic catalyst : Discoveries, Concepts and applications, (J. Wiley, New-York, 1983).
5. J. R. Anderson and M. Boudart, Catalysis by Metals and Alloys, *J. of Cat.* **165**, 285-286 (1997).
6. J. M. Thomas and W. J. Thomas, Principles and practice of heterogeneous catalysis (VCH, New-York, 1997).
7. G. L. Haller and W. M. H. Sachtler, Catalyst characterization: structure/function, *Cat. Today* **22**, 261-280 (1994).
8. J. I. Di Cosimo, C. R. Apesteguía, M. J. L. Ginés and E. Iglesia, Structural requirements and reaction pathways in condensation reactions of alcohols on Mg_yAlO_x Catalysts, *J. of Cat.* **190**, 261-275 (2000).
9. F. Garin and G. Maire, Correlations between the surface structure of bimetallic or alloy catalysts and hydrocarbon skeletal rearrangement mechanisms: approach to the nature of the active sites and to the reaction intermediates, *J. of Molecular Cat.* **52**, 147-167 (1989).
10. D. C. Koningsberger and B. L. Mojet, "Role and contributions of XAFS spectroscopy   in catalysis", *Topics in Cat.* **10** (2000).
11. K. J. Chao and A. C. Wei, Characterization of heterogeneous catalysts by XAS, *J. of El. Spec. and Related Phenomena* **119**, 175-184 (2001).
12. T. Shido and R. Prins, Application of synchrotron radiation to in situ characterization   of catalysts, *Current Opinion in Solid State and Materials Science* **3**, 330-335 (1998).
13. J. Lynch, Development of Structural Characterisation Tools for Catalysts, *Oil and Gas Science and Technology – Rev. IFP* **57**, 281-305 (2002).
14. D. E. Sayers, F. W. Lytle, and E. A. Stern, in Advances in X-ray Analysis, edited by B. L. Henke, J. B. Newkirk, and G. R. Mallett (Plenum, New York, 1970).
15. D. Bazin, L. Guczi, and J. Lynch, AWAXS in heterogeneous catalysis, *Applied Cat. A* **226**, 87-113 (2002).
16. D. Bazin, D. Sayers, and J. Rehr, Comparison between XAS, AWAXS, ASAXS, and Dafs techniques applied to nanometer-scale metallic clusters, *J. Phys. Chem.* **101**, 11040-11050 (1997).
17. B. S. Clausen, L. Grabaek, H. Topsoe, L. B. Hansen, P. Stoltze, J. K. Norskov, and   O.   H. Nielsen, A new procedure for particle size determination by EXAFS based on molecular dynamics simulations, *J. of Cat.* **141**, 368-379 (1993).
18. J. Barbier, Deactivation of reforming catalysts by coking - a review, *Applied Cat.* **23**, 225-243 (1986).
19. F. Garin, Mechanism of NO_x decomposition, *Applied Cat. A* **222**, 183-219 (2001).
20. L. Guczi, F. Solymosi, and P. Tétényi (Eds.), Proceedings of the 10th International Congress on Catalysis, Akadémiai Kiadó, Budapest, 1993.

21. A. Bensaddik, A. Caballero, D. Bazin, H. Dexpert, B. Didillon, and J. Lynch, *In situ* study by XAS of the sulfuration of industrial catalysts : the Pt and PtRe/Al$_2$O$_3$ system, *Applied Cat. A*, **162**, 171-180 (1997).

22. R. Revel, D. Bazin, E. Elkaim, Y. Kihn, and H. Dexpert, An in situ study using AWAXS and XAS of the catalytic system ZnAl$_2$O$_4$ supported on alumina, *J. Phys. Chem. B*. **104**, 9828-9835 (2000).

23. D. Bazin, I. Kovács, L. Guczi, P. Parent, C. Laffon, F. De Groot, O. Ducreux, and J. Lynch, Genesis of Co/SiO$_2$ catalysts : XAS study at the CoL$_{III,II}$ absorption edges, *J. of Cat.* **189**, 456-462 (2000).

24. C. J. Jacobsen, S. Dahl, B.S. Clausen, S. Bahn, A. Logadottir, and J.K. Nørskov, Catalyst design by interpolation in the periodic table: bimetallic ammonia synthesis catalysts, *J. Am. Chem. So.c* **123**, 8404-8405 (2001).

25. M. C. Desjonqueres and D. Spanjaard, Concept in surface physics, (Springer-Verlag, Berlin,1998).

26. S. Schneider, D. Bazin, F. Garin, G. Maire, M. Capelle, G. Meunier, and R. Noirot, NO reaction over nanometer scale platinum clusters deposited on γ-alumina: an XAS study, *Applied Cat. A* **189**, 139-145 (1999).

27. D. Bazin, Solid state concepts to understand catalysis using nanoscale metallic particles, *Topics in Cat.* **18** , 79-84 (2002).

28. D. Bazin and L. Guczi, New opportunities to understand the properties on nanomaterials through Synchrotron Radiation and theoretical calculations, *Recent Res. Dev. Phys. Chem.* **3**, 387-418 (1999).

29. D. Bazin, H. Dexpert, and P. Lagarde, Characterization of heterogeneous catalysts: The EXAFS tool, *Topics in Current Chemistry* **145**, 69-80 (1988).

30. P. Esteban, J. C. Conessa, H. Dexpert, and D. Bazin, Spectroscopic analysis of heterogeneous catalysts, in J. L. G. Fierro (Ed.), Elsevier, Amsterdam, 1992.

31. D. Bazin, D. Sayers, Comparison between XAS & AWAXS applied to nanometer scale supported monometallic clusters, *Jpn J. Appl. Phys.* **32-2**, 249-251 (1993).

32. J. Moonen, J. Slot, L. Lefferts, D. Bazin, and H. Dexpert, The influence of polydispersity and inhomogeneity on EXAFS of bimetallic catalysts, *Physica B: Condensed Matter* **208-209**, 689-690 (1995).

33. D. Bazin, L. Guczi, and J. Lynch, Real time in situ XANES approach to characterise electronic state of nanometer-scale entities, *Rec. Res. Dev. Phys. Chem.* **4**, 259-289 (2000).

34. M.S.P. Francisco, V.R. Mastelaro, A.O. Florentino, and D. Bazin, Structural study of copper oxide supported on a ceria-modified titania catalyst system, *Topics in Cat.* **18**, 105-111 (2002).

35. D. Bazin, A. Triconnet, and P. Moureaux, An EXAFS characterisation of the highly dispersed bimetallic PtPd catalytic system, *NIMB* **97**, 41-43 (1995).

36. A. Khodakov, N. Barbouth, J. Oudar, F. Villain, D. Bazin, H. Dexpert, and P. Schulz, Investigation of dispersion and Localization of Platinum Species in Mazzite Using EXAFS, *J. Phys. Chem. B* **101**, 766-770 (1997).

37. C. Micheaud-Especel, D. Bazin, M. Guérin, P. Marécot, and J. Barbier, Study of supported bimetallic Pd-Pt catalysts. Characterization and catalytic activity for toluene hydrogenation, *Reaction Kinetics and Cat. Letters* **69**, 209-216 (2000).

38. D. Bazin, A. Bensaddik, V. Briois, and Ph. Saintavict, M.S. calculation of XAS and XRD calculations: Application to nanometer scale Cu metallic particle, *J. de Phys.* C4, 481-485 (1996).

39. M. Fernandez-Garcia, C. Marquez Alvarez, and G. L. Haller, XANES-TPR Study of Cu-Pd Bimetallic Catalysts: Application of Factor Analysis, *J. Phys. Chem.* **99**, 12565-12569 (1995).

40. D. Bazin and J. J. Rehr, Limits and advantages of XANES for nanometer-scale metallic clusters, accepted in J. Phys. Chem.

41. C. Geantet , Y. Soldo, C. Glasson, N. Matsubayashi, M. Lacroix, O. Proux, O. Ulrich, and J. L. Hazemann, In situ QEXAFS investigation at Co K-edge of the sulfidation of a CoMo/Al$_2$O$_3$ hydrotreating catalyst, *Cat. Letter* **73**, 95-98 (2001).

42. J.-D. Grunwaldt, C.Keresszegi, T. Mallat, and A. Baiker, In situ EXAFS study of Pd/Al$_2$O$_3$ during aerobic oxidation of cinnamyl alcohol in an organic solvent, J. of Cat., In Press,

43. D. Bazin, D. Sayers, J. J. Rehr, and C. Mottet, Numerical simulation of the Pt L$_{III}$ edge white line relative to nanometer scale clusters, J. of Phys. Chem. 101, 5332-5336 (1997).

44. D. Bazin and L. Guczi, Soft X-ray absorption spectroscopy and heterogeneous catalysis, Applied Cat. General A 213, 147-162 (2001).

45. D. Bazin, L. Guczi, and J. Lynch, AWAXS in heterogeneous catalysis, Applied Cat. A **226**, 87-113 (2002).

46. W. Vogel, B. Rosner, and B. Tesche, Structural investigations of Au$_{55}$ organometallic complexes by X-ray powder diffraction and transmission electron microscopie, J. Phys. Chem. 97, 11611-11616 (1993).

47. N. S. Guyot-Sionnest, F. Villain, D. Bazin, H. Dexpert, F. Lepeltier, J. Lynch, and J. P. Bournonville, In situ EXAFS studies under high tempertaure and pressure of a Pt/Al$_2$O$_3$ catalyst during reduction and hydrocarbon conversion "X-ray Absorption Fine structure", Ed. S. Samar Hasnain, Ed. Elis Horwood, 493-495, 1991.

48. R. Revel, D. Bazin, A. Seigneurin, P. Barthe, J. M. Dubuisson, T. Decamps, H. Sonneville, J. J. Poher, F. Maire and P. Lefrancois, A new X-ray absorption experimental set-up dedicated to DeNOx catalysis, NIM B **155**, 183-188 (1999).

49. S. Schneider, D. Bazin, J. M. Dubuisson, M. Ribbens, H. Sonneville, G. Meunier, F. Garin, G. Maire, and H. Dexpert, Development of a reactive cell for EXAFS in situ studies of automotive postcombustion catalytic converters, J. of X-Ray Science and Technology **8**, 221-230 (2000).

50. D. Bazin, H. Dexpert, P. Lagarde, and J. P. Bournonville, XAS studies of bimetallic Pt-Re(Rh)/Al$_2$O$_3$ catalysts in the first stages of preparation, J. of Cat. **110**, 209-215 (1988).

51. D. Bazin, H. Dexpert, J. P. Bournonville, and J. Lynch, Bimetallic reforming catalysts: XAS of the particle growing process during the reduction, J. of Cat. **123**, 86-95 (1990).

52. D. Bazin, H. Dexpert, N. S. Guyot-Sionnest, J. P. Bournonville, and J. Lynch, EXAFS characterization of reforming catalysts : examples of recent applications, J. de Chim. Phys. **T86**, 1707-1713 (1989).

53. N. S. Guyot-Sionnest, F. Villain, D. Bazin, H. Dexpert, J. Lynch, and F. Lepeltier, In situ XAS studies under high temperature and pressure of a Pt/Al$_2$O$_3$ catalyst during reduction and hydrocarbon conversion, Part I. Cat. Letters **8**, 283-296 (1991).

54. A. Bensaddik, H. Dexpert, D. Bazin, F. Villain, B. Didillon, and J. Lynch, In Situ study by XAS of the sulfuration of the Pt and PtRe/Al$_2$O$_3$ systems,Physica B **208**, 677-678 (1995).

55. R. Revel, D. Bazin, B. Bouchet-Fabre, A. Seigneurin, and Y. Kihn, An in situ study using AWAXS and XAS of the binary metal oxide catalytic system SnO$_2$-ZnAl$_2$O$_4$ supported on alumina, J. de Physique IV, **12**, 309-322 (2002).

56. R. Revel, P. Parent, C. Laffon, and D. Bazin, NO adsorption on ZnAl$_2$O$_4$/Al$_2$O$_3$ powder: A NEXAFS study at the nitrogen K edge, Cat. Letter **74**, 189-192 (2001).

57. A. Yu. Khodakov, J. Lynch, D. Bazin, B. Rebours, N. Zanier, B. Moisson, and P. Chaumette, Reducibility of cobalt species in silica-supported Fischer-Tropsch catalysts, J. of Cat. **168**, 16-25 (1997).

58. L. Guczi, L. Borkó, Z. Schay, D. Bazin and F. Mizukami, CO hydrogenation and methane activation over Pd-Co/SiO$_2$ catalysts prepared by sol/gel method, Cat. Today **65**, 51-57 (2001).

59. L. Guczi, D. Bazin, Z. Schay, and I. Kovács, Nanoscale bimetallic catalysts: are they really bimetallic? Surface and Interface Analysis **34**, 72-75 (2002).

60. L. Guczi, D. Bazin, I. Kovacs, L. Borko, Z.Schay, J. Lynch, P. Parent, C. Lafon, G. Stefler, Zs. Koppany and I.Sajo, Structure of PtCo/Al$_2$O$_3$ and PtCo/NaY bimetallic catalysts : characterization by in situ EXAFS, TPR, XPS and by activity in CO hydrogenation, Topics in Cat. **20**, 129-139 (2002).

61. D. Bazin, L. Borkó, Zs. Koppány, I. Kovács, G. Stefler; L.I. Sajó, Z. Schay, and L. Guczi, Re-Co/NaY and Re-Co/Al$_2$O$_3$, Bimetallic Catalysts: in situ EXAFS study and catalytic activity, Cat. Letters **84**, 169-182 (2002).

62. F. Maire, M. Capelle, G. Meunier, J. F. Beziau, D. Bazin, F. Garin, J. L. Schmitt, and G. Maire, An XAS spectroscopy (XANES-EXAFS) Investigated of PtRh automotive catalysts, *Preprints CAPOC III*, **1**, 161-166 (1994).
63. M. Boudart, Model catalysts: reductionism for understanding, *Topics in Cat.* **13**, 147-149 (2000).
64. S. Schneider, S. Ringler, P. Girard, G. Maire, F. Garin, and D. Bazin, Mechanistic studies of the NO_x reduction by hydrocarbon in oxidative atmosphere, *Stud. Surf. Sci. Catal.* **130**, 641-645 (2000).
65. S. Schneider, D. Bazin, F. Garin, G. Maire, M. Capelle, G. Meunier, and R. Noirot, NO reaction over nanometer scale platinum clusters deposited on γ-alumina: an XAS study, *Applied Cat. A* **189**, 139-145 (1999).
66. P. Loof, B. Stenbom, H. Norden, and B. Kasemo, Rapid sintering in NO of nanometer-sized Pt particles on γ-Al_2O_3 observed by CO temperature-programmed desorption and transmission electron microscopy, *J. of Cat.* **144**, 60-76 (1993).
67. W. Brown, and D. A. King, NO Chemisorption and Reactions on Metal Surfaces: A New Perspective, *J. Phys. Chem. B.* **104**, 2578-2595 (2000).
68. W. Brown, and D. A. King, NO Chemisorption and Reactions on Metal Surfaces: A New Perspective, *J. Phys. Chem. B.* (Addition/Correction) **104**, 11440-11440 (2000).
69. D. Bazin, Solid state concepts to understand catalysis using nanoscale metallic particles, *Topics in Cat.* **18** , 79-84 (2002).
70. J. Evans and M. A. Newton, Towards a structure/activity relationship for oxide supported metals, *J. of Molecular Cat. A*, **182-183**, 351-357 (2002).
71. T. Campbell, A. J. Dent, S. Diaz-Moreno, J. Evans, S. G. Fiddy, M. A. Newton, S. Turin, Susceptibility of a heterogeneous catalyst, Rh/Al_2O_3, to rapid structural change by exposure to NO, *Chem. Commun.* **4**, 304-305 (2002).
72. S. Schneider, D. Bazin, G. Meunier, R. Noirot, M. Capelle, F. Garin, G. Maire, and H. Dexpert, An EXAFS study of the interaction of different reactant gases over nanometer scale Pt clusters deposited on γ-Al_2O_3, *Cat. Letters* **71**, 155-162 (2001).
73. A. Baiker, Progress in asymmetric heterogeneous catalysis: Design of novelchirally modified platinum metal catalysts, *J. Mol. Catal. A* **115**, 473-493 (1997).
74. Z. Kónya, V. F. Puntes, I. Kiricsi, J. Zhu, P. Alivisatos, and G. A. Somorjai, Novel Two-Step Synthesis of Controlled Size and Shape Platinum Nanoparticles Encapsulated in Mesoporous Silica, *Cat. Letters* **81**, 137-140 (2002).
75. G. A. Somorjai, New model catalysts (platinum nanoparticles) and new techniques (SFG and STM) for studies of reaction intermediates and surface restructuring at high pressures during catalytic reactions, *Applied Surface Science* **121-122**, 1-19 (1997).

21

Design, Building, Characterization and Performance at the Nanometric Scale of Heterogeneous Catalysts

J.-M. Basset,[*] J.-P. Candy, C. Copéret, F. Lefebvre, and E. A. Quadrelli

21.1. INTRODUCTION

Whether it is **homogeneous** or **heterogeneous** (or even enzymatic), catalysis is *primarily a molecular phenomenon* since it involves the chemical transformation of molecules into other molecules. At the beginning of the 21st century, even if many attempts have been made to fully erase the gap existing between these two important fields of chemistry, they still belong to different scientific communities : homogeneous catalysis is related to molecular organometallic chemistry, and heterogeneous catalysis is closer to surface science and solid state chemistry.

Although the physico-chemical tools of surface science have progressed tremendously in the last decades, the level of understanding of heterogeneous catalysis is still limited, especially when compared to that of molecular organometallic chemistry and homogeneous catalysis. One of the main reasons for the difficulty in obtaining a « structure-activity » relationship in heterogeneous catalysis may be due to the small number of active sites and their dispersion.

It is to meet this challenge that *Surface Organometallic Chemistry* (SOMC) has been developed. This discipline of chemistry brings the concepts and the tools of molecular chemistry, especially organometallic chemistry, to surface science and heterogeneous catalysis.

The ultimate goal of this chemistry is to construct on surfaces, via a truly molecular engineering approach, the active sites, uniform in composition and distribution, so as to

---

[*]Laboratoire de Chimie Organométallique de Surface, UMR 9986 CNRS – CPE Lyon. 43 Bd du 11 Novembre 1918 F-69616 Villeurbanne Cedex, France.

achieve « single-site » heterogeneous catalysts. From this point of view, SOMC can be understood as a nanotechnology as its purpose is to work on atomic objects and to modify them in order to obtain, over all the surface of a solid the same catalytic species with a structure known as perfectly as possible (how the metal is bound to the solid, what are its ligands, …). As a consequence, and the examples described below will show it, the synthesis and characterization of a catalyst will be described by reactions involving molecules whose size is smaller than 1 nm and which will cover objects having sizes higher by two or more orders of magnitude. Note that pioneering work started in the late 60's by polymer scientists like D. Ballard at ICI and Y. Yermakov at the Novosibirsk Institute of Catalysis,[1-6] but this field of chemistry in its present form, with a precise definition of the overall structure of the active sites, has just recently emerged via the development of more performing analytical tools.

One of the first aspects of this chemistry has been to study the reactivity of organometallic molecules (of main group elements, transition metals, lanthanides and actinides) with surfaces such as oxides (silica, alumina, zeolites, mesoporous materials) or metal surfaces (in the sense of metallic materials), to develop performing nanostructured catalysts. This chapter will be concerned with selected examples in this field. In particular, after a brief overview on the techniques and tools in SOMC with respect to the solution analogues (§ 2); selected catalysts developed by SOMC on amorphous silica will be presented (§ 3), with applications to olefin metathesis, olefin polymerisation, and low temperature alkane transformation; the second part of the section (§ 3.2) will be devoted to the advancements in the field linked to a nano-scale control of the surface (in contrast to the amorphousness of the previous support) that can be gained with zeolites and mesoporous materials. Finally, the application of SOMC to the modification of nano-structured metallic surfaces with mononuclear organometallic compounds is addressed (§ 4).

21.2. TECHNIQUES AND TOOLS IN SURFACE ORGANOMETALLIC CHEMISTRY

One of the main objective of *Surface Organometallic Chemistry* is to be able to characterise surface species in order to establish structure/activity relationships as in homogeneous catalysis. Therefore, as in molecular chemistry, *Surface Organometallic Chemistry* heavily relies on both chemical and spectroscopic methods to understand the structure of surface entities. The most powerful and commonly used tools in structural molecular organometallic chemistry are X-Ray crystallography, NMR, IR (Raman), and ESR spectroscopies which give a clear and detailed picture of the coordination sphere around a metal centre.

In this respect, *Surface Organometallic Chemistry* has therefore used the corresponding methods for amorphous solids (Scheme 21.1, next page) : (i) EXAFS, which provides distances as well as average coordination numbers (the number of first neighbouring atoms, and in some cases second ones), (ii) IR (usually *in situ*), which is used to follow modification of both the support (especially during grafting) and the surface complex upon further treatment, (iii) ESR and XANES, which can point out the oxidation state and the geometry of the metal complex when applicable, and (iv) solid state NMR (1D and 2D), probably one of the most powerful methods for structure determination. Moreover, since *Surface Organometallic Chemistry* deals with surfaces, it also uses other important methods of surface science such as XPS and specific surface measurements (BET, Porous distribution).[7]

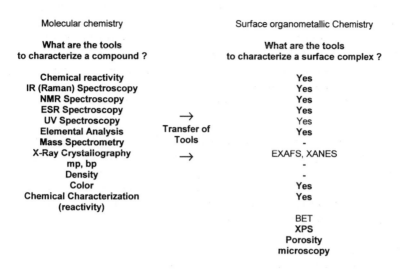

Molecular chemistry		Surface organometallic Chemistry
What are the tools to characterize a compound ?		**What are the tools to characterize a surface complex ?**
Chemical reactivity		Yes
IR (Raman) Spectroscopy		Yes
NMR Spectroscopy		Yes
ESR Spectroscopy		Yes
UV Spectroscopy	→	Yes
Elemental Analysis	Transfer of	Yes
Mass Spectrometry	Tools	-
X-Ray Crystallography	→	EXAFS, XANES
mp, bp		-
Density		-
Color		Yes
Chemical Characterization (reactivity)		Yes
		BET
		XPS
		Porosity
		microscopy

SCHEME 21.1

IR Spectroscopy is an essential tool for understanding the grafting step and to assess the modification of the species on the surface upon various treatments (chemical, photochemical and thermal reactivity). This, in combination with quantitative measurements of product evolution during grafting and upon further chemical treatments of the surface complex, can provide a quick, yet clear, picture of the surface complex via mass balance analysis. Typically, the latter aspect has been mainly addressed by using reactions of hydrolysis, alcoholysis, hydrogenolysis, oxidation and in some specific cases pseudo-Wittig reactions. Additionally grafting on deuterated silica ($\equiv$Si-O-D) or using deuterolysis of surface complexes can provide more information about the mode of grafting and the coordination sphere of the grafted organometallic complex.

All these stoichiometric events and spectroscopic methods help to understand the structure of a surface complex. Several recent examples coming from our laboratory, that illustrate the approach, will be presented including their relevance to catalysis and to well-defined materials.

21.3. SURFACE ORGANOMETALLIC CHEMISTRY ON OXIDES

21.3.1. Surface Organometallic Complexes on Silica

A key component in *Surface Organometallic Chemistry* on oxide is the support. It is therefore of prime importance to control its reactivity (selection of reactive functional groups). For example, on a silica 200 m^2/g Aerosil® from Degussa, the surface of the support (which is made of globular particles having *ca.* 15 nm in diameter) is composed of siloxane bridges ($\equiv$Si-O-Si$\equiv$) and silanol groups ($\equiv$Si-OH). The functional groups are either silanols or siloxane bridges, and their concentrations depend highly on the temperature of partial dehydroxylation of silica. The $\equiv$Si-O-Si$\equiv$ groups which are formed by dehydroxylation of the support at high temperatures are believed to be rather dispersed in terms of bonding energy. As a result, the surface organometallic chemistry on siloxane bridges is still under developed. On the contrary, despite the amorphous structure of silica , the types of reactive functional groups ($\equiv$SiOH) and their pK_a are rather homogeneous in this support compared to others.[7-11] Therefore, the surface organometallic chemistry on silica relies highly on the reactivity at the surface silanol moieties.

21.3.1.1. Well-Defined Surface Organometallic Complexes. The characterisation of a surface complex requires the combination of both chemical analysis and spectroscopic studies. The reaction of [Ta($=$CH*t*Bu)(CH$_2$*t*Bu)$_3$] with a SiO$_{2-(700)}$ (silica partially dehydroxylated at 700 °C) to yield the monosiloxy surface complex [($\equiv$SiO)Ta($=$CH*t*Bu)(CH$_2$*t*Bu)$_2$], (**1**),[12, 13] will be presented hereafter in details to illustrate the investigative tools portfolio used routinely to assess the nature of the surface organometallic complexes.

Upon reaction of the surface with the starting organometallic complex [Ta($=$CH*t*Bu)(CH$_2$*t*Bu)$_3$], the *in situ* IR study shows the disappearance of the free surface silanol stretches (υ(OH) at 3747 cm^{-1}) , and the appearance of alkyl stretching and bending vibrations (υ(CH) at 3000-2800 cm^{-1} and δ(CH) at 1500-1300 cm^{-1}) due to surface-bound metallo-organic species. Upon reaction, GC-mass analyses of the evolved gases show the production of 1.0 eq. NpH/grafted Ta (NpH = neopentane) . The elemental analyses of the material yield a Ta and C weight content of 4.25 and 3.42 %, respectively, for a C/Ta ratio of 12, consistent with the proposed structure for **1**. The 1D solid state NMR spectroscopy on the partially ^{13}C-enriched [$\equiv$SiO-Ta($=$*CH*t*Bu)(*CH$_2$*t*Bu)$_2$] at the α-position of the metal centre clearly points out the presence of neopentyl and neopentylidene ligands.[14] Furthermore, 2D HETCOR solid state NMR is a key technique to ascertain the identity of surface complexes : for example, it confirms the correlation between the carbenic proton (^{1}H NMR peak at 4.2 ppm) and carbon (^{13}C NMR peak at 246 ppm) in **1**. Moreover, EXAFS data are also fully consistent with the structure of **1**, giving four neighbouring atoms around the Ta centre at 2.15 Å (2 Ta-C) and 1.89 Å (1 Ta-O + 1 Ta$=$C) along with a siloxane bridge at 2.64 Å.[15] Finally, the chemical composition of the surface organometallic complex is further assessed by reactivity studies on the functionalised material. Namely, methanolysis of **1** leads to the evolution of 2.90 eq. NpH/Ta, consistently with the presence of three alkyl moieties in the metal coordination sphere.

Therefore, all the observations, chemical analysis and spectroscopic studies (*viz. in situ* IR, elemental analyses, mass balance analysis, 1D and 2D NMR, EXAFS, chemical reactivity, ...) are consistent with Eq. (1) and allow the claim that the precise molecular nature of surface organometallic complex **1** is known.

$$[Ta(=CH\textit{t}Bu)(CH_2\textit{t}Bu)_3] + \equiv SiOH \rightarrow [\equiv SiO\text{-}Ta(=CH\textit{t}Bu)(CH_2\textit{t}Bu)_2] + NpH \quad (1)$$
$$SiO_{2-(700)} \qquad\qquad\qquad\qquad \mathbf{1}$$

The surface of the silica particles will then be covered (with a density of 0.7 complex per nm^2) by this monografted complex whose projected area is 0.9 nm^2, hence achieving a surface coverage of about 90%.

Furthermore, it has been shown that it is possible to consider the surface as a ligand in its own right by controlling the concentration of the surface functional groups of silica. For example, the use of a SiO$_{2-(300)}$ rather than SiO$_{2-(700)}$ with [Ta(=CHtBu)(CH$_2t$Bu)$_3$] yields a material supporting a bis(siloxy) surface complex, [(≡SiO)$_2$Ta(=CHtBu)(CH$_2t$Bu)] (**2**), at a higher density with respect to its SiO$_{2-(700)}$ analogue (*viz.* 1.2 *vs* 0.7 complexes/ nm^2).

In general, it has been shown that it is possible to control and characterize the formation of other supported transition metal complexes by the same procedure. For example in the context of neopentyl derivatives, the corresponding surface complexes of group 4, 6 and 7 transition metals, namely [≡SiO-M(CH$_2t$Bu)$_3$] (M = Ti, Zr or Hf), [≡SiO-M(≡CtBu)(CH$_2t$Bu)$_2$] (M = Mo, **4**;[16] M = W, **5**[15,17]) and [≡SiO-Re(≡CtBu)(=CHtBu)(CH$_2t$Bu)], **6**, (Scheme 21.2) have been synthesised and characterized by combining chemical analysis and spectroscopy (IR, EXAFS, 1D and 2D NMR) like for the characterisation of **1**.[18-20] The nanometric features of these materials are similar to the ones reported above for **1**, given the similar molecular environment of the grafted metallic centres.

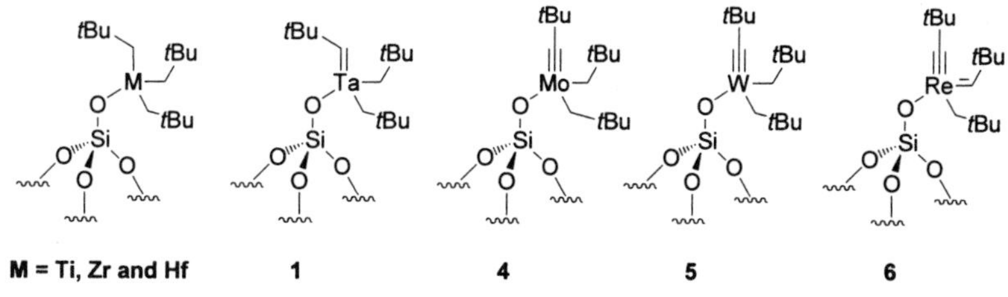

M = Ti, Zr and Hf **1** **4** **5** **6**

SCHEME 21.2

21.3.1.2. Molecular Analogues of Silica : Another Tool to Understand the Chemistry of Surfaces. Molecular chemistry is usually faster and easier in terms of structure characterisation and reactivity understanding. Therefore the use of molecular models to mimic the surface of silica can help in understanding surface reactions. For example, the reaction of organometallics with these models yields metal complexes that are more easily characterised and that can provide important data like bond distances (X-Ray and/or EXAFS) and chemical shifts (solution and/or solid state NMR), which can be used as feedback for data obtained for surface organometallic complexes (solid state NMR and EXAFS).[21-29]

For example, the study of **1** is facilitated by the use of a corresponding molecular complex [(C$_5$H$_9$)$_7$(Si$_7$O$_{12}$)SiO-Ta(=CHtBu)(CH$_2t$Bu)$_2$], (**1m**), prepared by the reaction of a molecular analogue of an isolated silanol of silica. Compound **1m** displays chemical shifts close to or identical in ^{1}H- and ^{13}C-NMR to those of **1** (Scheme 21.3, next page). This reaction, easily monitored by NMR spectroscopy has allowed the observation of a reaction intermediate **3m**, namely [(C$_5$H$_9$)$_7$(Si$_7$O$_{12}$)SiO-Ta(CH$_2t$Bu)$_4$], which corresponds to the addition product of the silanol to the carbenic moiety of [Ta(=CHtBu)(CH$_2t$Bu)$_3$] and which slowly decomposes via α-H abstraction into **1m** [(C$_5$H$_9$)$_7$(Si$_7$O$_{12}$)SiO-Ta(=CH*t*Bu)(CH$_2$*t*Bu)$_2$]. This led to the observation of the corresponding intermediate **3** on the silica surface by solid state NMR spectroscopy if low temperatures and short evaporation times were used during the reaction of [Ta(=CH*t*Bu)(CH$_2$*t*Bu)$_3$] with silica.

This example clearly shows the importance of combining several techniques in order to establish the structure of surface complexes and their mechanism of formation.

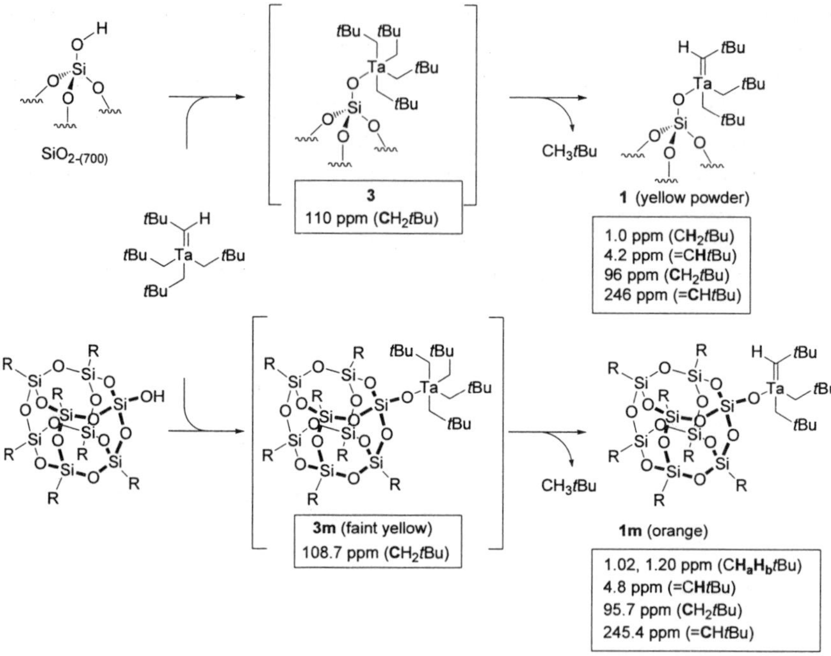

SCHEME 21.3

21.3.1.3. From Well-Defined Complexes to Highly Active Olefin Metathesis Catalysts. In the area of olefin metathesis, there is a need for well-defined metallocarbenes (or carbynes) in order to investigate the influence of various ligands on the catalytic properties of surface complexes. Along this line [(≡SiO)Ta(=CH*t*Bu)(CH$_2$*t*Bu)$_2$] (**1**) shows to no little activity in olefin metathesis. On the other hand the well-defined silica supported alkylidene Re-complex [(≡SiO)Re(≡C*t*Bu)(=CH*t*Bu)(CH$_2$*t*Bu)] (**6**) has displayed unprecedented activities in olefin metathesis[18-20, 30] compared to all the existing heterogeneous Re-based catalysts[31-37] and even homogeneous ones. [31-41] Moreover well-defined alkylidyne surface complexes **4** [(≡SiO)Mo(≡C*t*Bu)(CH$_2$*t*Bu)$_2$] and **5** [(≡SiO)W(≡C*t*Bu)(CH$_2$*t*Bu)$_2$] also show good activity in olefin metathesis even if the formation of the active species (alkylidene propagating centre) is not clear in these cases.[15, 17]

These catalysts, based on supported carbene or carbyne complexes of Mo, W or Re, exhibit activities, selectivities and lifetimes in some cases superior to those encountered in classical heterogeneous and even homogeneous catalysis.

21.3.1.4. From Well-Defined Complexes to Olefin Polymerisation Catalysts. In homogeneous catalysis, one of the major developments in olefin polymerisation was the discovery of the metallocene catalysts and the possible control of the micro-structure and properties of polymers by the right choice of ligands and co-catalysts, in particular MAO.[42, 43]

This immediately led to investigation of the synthesis of their corresponding well-defined supported surface complexes like **7** [(≡SiO)ZrCp$_2$Me] and **8** [(≡SiO)ZrCp*Me$_2$], which have been fully characterised.[44] However, these grafted systems show little activity in polymerisation. Grafting these metallocenes on supports containing Lewis acid centres such as silica-alumina (SiO$_2$/Al$_2$O$_3$) or alumina (Al$_2$O$_3$) did improve the system since partially cationic Zr centres are formed via the transfer of a methyl ligand onto an adjacent vacant aluminium sites. The following order of activity has been found :

$$SiO_2\text{-}Al_2O_{3(500)} < Al_2O_{3(1000)} < Al_2O_{3(500)}$$

Nevertheless, the activity obtained for the best catalyst, Cp*ZrMe$_3$/Al$_2$O$_{3(500)}$, remained low by comparison to the level of activity reported for the homogeneous systems.

Recently EXXON Corp. has disclosed a MAO-free route, which relies upon the formation of well-defined cationic metal-alkyl surface species. The reaction of silica sequentially with B(C$_6$F$_5$)$_3$ and dimethylaniline yields [(≡SiO-B(C$_6$F$_5$)$_3$)⁻(HNMe$_2$Ph)$^+$],[45-48] a modified silica with siloxyborato anilinium surface functional groups (Scheme 21.4), which reacts with dialkyl transition-metal complexes to form the corresponding cationic systems. These systems display high activities in the polymerisation of ethylene and propylene.

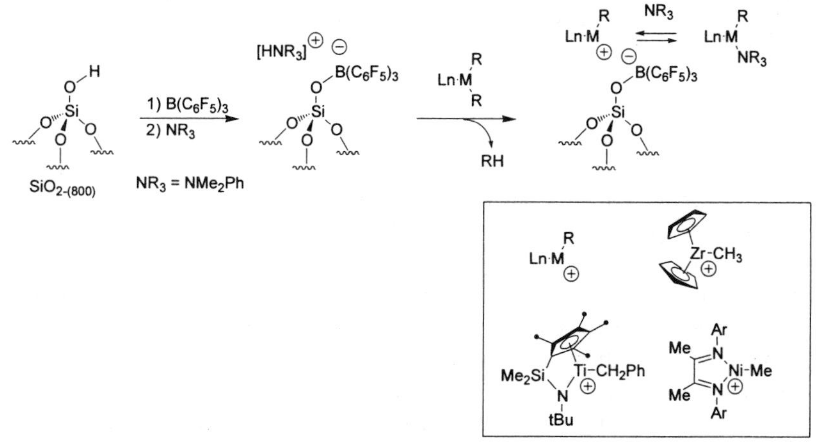

SCHEME 21.4

21.3.1.5. From Well-Defined Alkoxide Complexes to Highly Active and Selective Epoxidation Catalysts. In the homogeneous Sharpless epoxidation system, the proposed "active" species probably consists of a d^0-titanium centrer surrounded by the bidentate tartrate, an allyloxy (from the allylic alcohol) and a t-butylperoxy (from t-butylhydroperoxide, TBHP) ligands (Scheme 21.5).

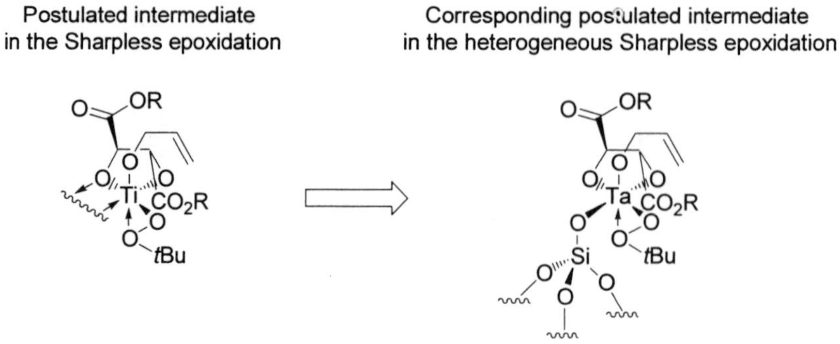

Postulated intermediate
in the Sharpless epoxidation

Corresponding postulated intermediate
in the heterogeneous Sharpless epoxidation

SCHEME 21.5

Grafting a Ti complex on a metal oxide support could not lead to the coordination sphere described before without losing one of these four key ligands (the C_2-symmetric chiral ligand, the oxidant and the olefin). In order to obtain a coordination sphere similar to that of the proposed Sharpless epoxidation active species, one more valence electron is required like in tantalum, a group V metal. In this case, tantalum can accommodate all ligands necessary for asymmetric epoxidation catalysis including that of the silica surface. Ethanolysis of **1** followed by a treatment with diethyl tartrate (DET) gives the corresponding $[(\equiv SiO)Ta(DET)(OEt)_2]$. This epoxidation catalyst usually displays activity and selectivity better than those obtained with the original Sharpless homogeneous catalyst, with the advantage of being easily separable from the reaction medium and recycled with no detectable leaching.[49]

21.3.1.6. From Well-Defined Hydride Complexes to a New Generation of Catalysts for Alkane and Polyolefin Transformation. Upon thermal treatment under H_2 of surface perhydrocarbyl complexes, it is possible to generate on oxide surfaces transition metal hydrides. Therefore $[(\equiv SiO)Ta(=CHtBu)(CH_2tBu)_2]$ and $[(\equiv SiO)M(CH_2tBu)_3]$ (M = Ti, Zr, Hf) are converted into $[(\equiv SiO)_2Ta-H]$ (**9**)[50] and $[(\equiv SiO)_3M-H]$[51-54] (**10**) respectively via hydrogenolysis of the perhydrocarbyl ligands and further reaction with the siloxane bridges to give the hydride and surface (Si-H) moieties (Scheme 21.6, next page).

The density of the grafted Ta-H moieties over the 15 nm silica particles is equal to that of their precursor, **1**. Obviously, given the smaller size of an hydrid ligand with respect to three *tert*-butyl containing ligand, the surface coverage in terms of non-accessible support area is much lower. The structure of **9** and **10** has been established by using both chemical reactivity studies and EXAFS. For example EXAFS data on **9** confirms the presence of 2 oxygen atoms at 1.89 Å from the Ta centre.

These hydride complexes readily react with alkanes to give stable surface alkyl complexes and H_2 and participate in H/D exchange processes.[55-57] More interestingly, these complexes also participate in catalytic reactions that involve a C-C bond activation as a key step. For example $[(\equiv SiO)_3Zr-H]$ readily transforms polyethylene into lower alkanes in the presence of hydrogen ; the depolymerisation process results in fact from successive β-alkyl transfer/hydrogenolysis steps.[58, 59] While [(≡SiO)$_2$Ta-H] also catalyses the hydrogenolysis of alkanes,[60] it transforms acyclic alkanes into their lower and higher homologues in the absence of H$_2$.[61-63] This new reaction has been named alkane σ-bond metathesis since alkyl fragments of alkane mixtures are exchanged in contrast with the well-known alkene metathesis, for which the alkylidene fragments are exchanged.

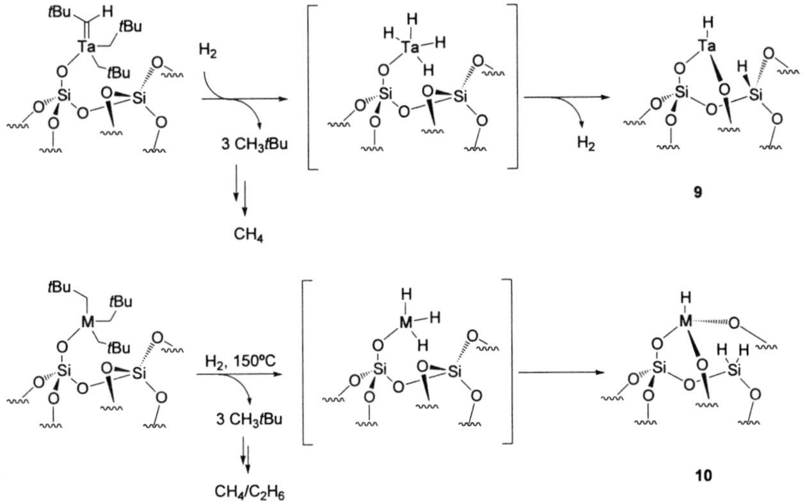

SCHEME 21.6

21.3.2. Surface Organometallic Complexes on Micro- and Mesoporous Materials: New Solids for Nano-catalysts and Nano-objects

Mesoporous silicas such as MCM-41 or HMS can be considered, in a first approximation, as other types of silicas and therefore the same strategy and methodology described above for silica can be applied to these materials. The main advantage of these materials is that they display a very large surface area (typically around 1000 m^2.g^{-1} while Aerosil silica is only 200 or 300 m^2.g^{-1}), present higher mesoscale order (typically hexagonal ordered channels of 0.3-0.8 nm diameter in MCM-41) and do not give strong inter-particular interactions. As a consequence, the metal loading can be increased considerably compared to classical supports (e.g. the MCM-41 analogue of 1 can contain up to 15% of grafted Ta, with respect to the typical 6% obtained for 1), allowing an easier characterization of the grafted organometallic fragments and increasing the catalytic efficiency.

The use of mesoporous silicas can also lead to unexpected properties, due to their morphological differences from classical fumed silica. The high surface area is mainly due to the presence of mesopores (from 2 to 10 nm in diameter) regular in both shape and position, resulting in quasi-crystalline materials. Grafting organometallic complexes in these pores will result in a transformation of the surface, initially hydrophilic, into a hydrophobic material, rendering the sorption properties of hydrocarbons completely different.[64, 65]

Zeolites and molecular sieves can lead to more complex reactions. Indeed even if these solids display a regular structure due to their crystallinity, grafting sites can be found inside the pores (in the channels and in the bulk of the crystals) and on the external surface, at the cut-off of the crystals. Both types of reactions can be observed by a judicious choice of the organometallic complex and the molecular sieves: when the organometallic complex is too big to enter the pores, it will react with the grafting sites present on the external surface. A typical example is the grafting of various tin fragments ($SnCy_3$, $SniPr_3$, $SnMe_3$) on the external surface of Mordenite.[66,67]   The resulting materials display very interesting properties in hydrocarbons separation (Figure 21.1): While hexane, 2-methylpentane, 2,3-dimethylbutane and isooctane readily enter the Mordenite cavity, some are selectively excluded when the size of the pore entrance has been tuned by the grafting of an organotin fragment. This discrimination depends on the organic group on tin : Cy > iPr > Me and the alkane. It is therefore possible to separate the hydrocarbons by using these modified zeolites as a solid phase in a gas chromatography.

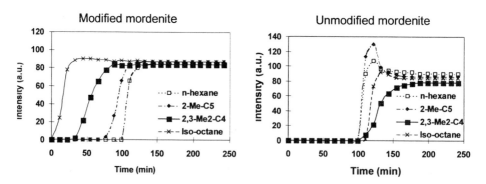

FIGURE 21.1. Separation of a mixture of hydrocarbons (n-hexane, 2-methyl pentane, 2,3-dimethyl butane and iso-octane) by unmodified mordenite and by mordenite modified by grafting tributyltin fragments

Another interesting aspect of the separation of organometallic complexes with hydroxyl groups on the external surface of zeolites is that they will "passivate" the solid's outer surface. Indeed, very often in catalysis, the selectivity induced by the limited size of the pores of zeolites is reduced or killed by secondary reactions which occur on the external surface. Suppressing these secondary reactions is then of primary importance and be achieved by reaction of the external surface with organometallic complexes. A typical example is the isomerisation on Mordenite (Scheme 21.7, next page).

If unmodified Mordenite is used, the selectivity to the desired product (para-xylene) is low and by-products (toluene and mesitylene) are obtained. If Mordenite has been passivated by its reaction with organometallic complexes, the isomerisation reaction can only occur in the Mordenite channels, suppressing bimolecular reactions responsible for the by-products. [68, 69]

The other possibility with zeolites or molecular sieves is to use organometallic complexes which can fill the pores of the material. In this case, the higher concentration of reactive sites inside the pores (about two order of magnitudes) leads to a principal grafting reaction in the channels and/or cages. Moreover, they usually display a higher reactivity. For example, while on silica the grafting reaction of tin complexes occurs only slowly at 150 °C, on Cloverite it occurs slowly at room temperature and rapidly at 150 °C.[70-72] On Faujasite the grafting occurs rapidly at room temperature and even at –50 °C.[65, 73] Due to the presence of the transition metal in the channels, these materials should present particular catalytic properties that are currently under investigation.

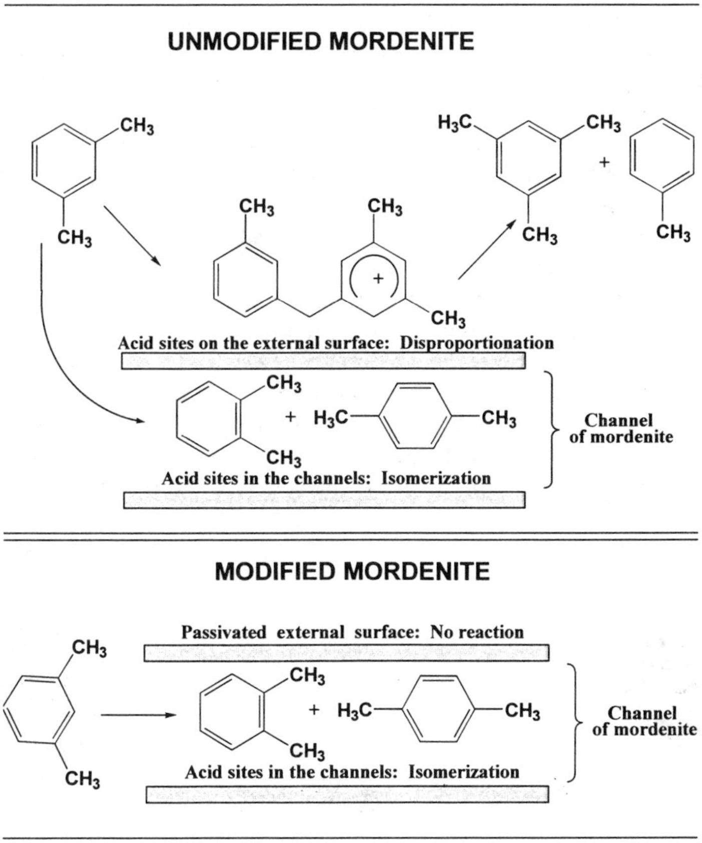

SCHEME 21.7

This field is emerging in parallel with the preparation of new well-defined solids and should provide a way to tune and control the properties of these materials.

21.4. SURFACE ORGANOMETALLIC CHEMISTRY ON METALS

21.4.1. Generalities on Surface Organometallic Chemistry on Metals

This field is devoted to the understanding of the reactivity of organometallic reagents with metal surface in order to generate well-defined bimetallic nano-structured catalysts or materials, and will use the same tools as previously exemplified. Its development has been made possible for two main reasons : firstly, it is possible to prepare very small metal particles (with a nanometer size) highly dispersed on a support with a narrow particle-size distribution. Secondly, when a metallic particle of a zero valent metal is grafted on an oxide support, most organometallic compounds have a higher chemical affinity with the particle than with the oxide support. These two parameters render the field of surface organometallic chemistry on metals possible even if the surface area of a highly dispersed Pt/SiO_2, for example, does not exceed a few square meters compared with the surface area of the oxide support (several hundred square meters). This nanoscience can then be described as the synthesis of subnanometer objects (the grafted organometallic fragments) on nanometer objects (the metal particles) which are themselves deposited on supports having a size two orders of magnitude higher.

Herein the synthesis, the characterisation and the use of bimetallic tin containing catalysts will be presented. The controlled reaction (below 50 °C) of Bu_4Sn with the surface of a metal particles (M_S), characterised by electron microscopy and chemisorption methods (hydrogen, oxygen and carbon monoxide adsorption) generates M_S-$SnBu_3$ via cleavage of one Sn-C bond.

Naturally, depending on the metal and the reaction conditions, the metal-grafted tributyltin surface complex can further evolve towards dibutyl, monobutyl and even totally de-alkylated tin species (Scheme 21.8). In the latter case, the tin atom can be considered as an adatom on the metal surface. By thermal treatment, this adatom migrates into the first layers of the metal particle, leading to the formation of a surface alloy. Using judicious reaction conditions, it is possible to obtain for a given metal a single and well-defined environment for tin, which can be fully characterised by physico-chemical methods including EXAFS.

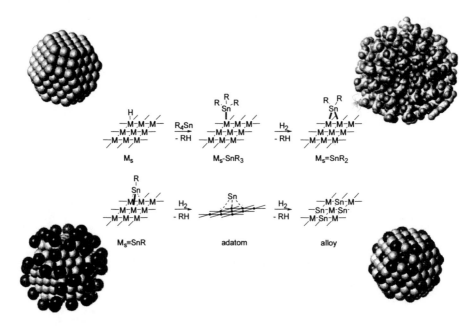

SCHEME 21.8. Various steps observed during the hydrogenolysis of $SnBu_4$ at the surface of a Pt particle (sphere color-code: Pt light grey, H grey, C white, Sn dark grey).

For instance, the monografted organotin species (M_S-SnBu$_3$) can be obtained by reaction of SnBu$_4$ with a Ni/SiO$_2$ catalyst covered with chemisorbed hydrogen. Its characterisation by EXAFS gave one Sn-Ni bond at 2.68 Å and three Sn-C bonds at 2.17 Å in the first coordination sphere of tin.[74] The bis(grafted) organotin species [(M_S)$_2$SnBu$_2$] can be prepared on rhodium particles supported on silica[75] for which the EXAFS studies gave two Sn-Rh bonds at 2.62 Å and two Sn-C bonds at 2.12 Å. The tris(grafted) tin surface complex [(M_S)$_3$SnBu] was isolated on a Pt/SiO$_2$ catalyst.[76] EXAFS measurements indicate that tin is surrounded by three platinum atoms at 2.68 Å and one carbon atom at 2.11 Å. Naked tin adatoms can also be obtained on a Pt/SiO$_2$ catalyst but, in order to remove all butyl ligands, the solid needs to be further heated to 300 °C.[77] EXAFS data indicate that tin is only surrounded by four platinum atoms at the same distance of 2.76 Å, which clearly indicates that tin is located on the metal surface and not in the bulk. Further heating the above sample under hydrogen at 500 °C results in an increase in the number of platinum atoms surrounding tin up to ca. 5, which can be explained by a migration of the tin atom into the first platinum layer.

Depending on the surface arrangements described above, it will be possible to tune the catalyst/material for a specific property.

21.4.2. Application of Surface Organometallic Chemistry on Metals to Catalysis

When the hydrogenation of citral is performed with a supported metal, for example Rh/SiO$_2$ under classical conditions (liquid phase, rhodium particles dimensions 1.8 nm, rhodium dispersion 80 %, citral/Rh$_S$ = 200, P(H$_2$) = 80 bars, T = 340 K) the catalytic activity is very high but the reaction is not selective. On the other hand, when Rh/SiO$_2$ is modified by reaction with SnBu$_4$ (Sn/Rh$_S$ = 0.95) to give [(Rh$_S$)$_2$SnBu$_2$] as a major surface species, the catalytic activity is only slightly decreased while 3,7-dimethyl-2, 6-octenol is obtained with selectivity higher than 95 %, see Eq. (2).[78]

In the case of adatoms, their formation induces a poisoning of undesirable sites, resulting in a significant increase or change of the selectivities. For example, for a Sn/Rh$_S$ ratio of 0.15, the hydrogenation of citral is very selective (75%) in the formation of citronellal, see Eq. (3).[79]

Rh-SnBu$_2$/SiO$_2$
Rh$_S$/Sn = 0.95

3,7-dimethyl-2,6-octenol
sel. > 95 %

(2)

citral

Rh-Sn/SiO$_2$
Rh$_S$/Sn = 0.15

citronellal
sel. = 75 %

(3)

In the case of alloys, there is a site isolation effect. For instance, the dehydrogenation of isobutane into isobutene proceeds at high temperature (ca. 550 °C) and low hydrogen pressure (1 bar) in the presence of silica-supported platinum particles of about 2 nm in size.

Under these conditions, the catalyst is very active (turnover frequency 5 s^{-1}), but the selectivity is only 93 % since by-products are formed due to the hydrogenolysis properties of the metal. On the other hand, using a Sn/Pt alloy.[76, 79] the catalyst becomes totally selective for isobutene (for example, when the Sn/Pt$_S$ ratio is equal to 0.85, the selectivity to isobutene is 99.5 %). This increase in selectivity is due to the bimetallic nature of the catalysts, for which Pt atoms are isolated by Sn atoms (site isolation), Eq. (4).

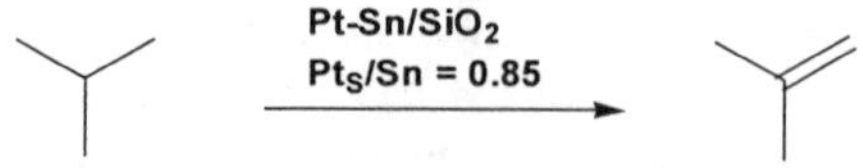

$$\text{Pt-Sn/SiO}_2 \qquad \text{Pt}_S/\text{Sn} = 0.85 \tag{4}$$

While β-H elimination, the key step for the formation of isobutene, involves one metal center, hydrogenolysis, responsible for the formation of by-products, requires two metal centers. Therefore isolating the active sites increases the selectivity (Scheme 21.9).

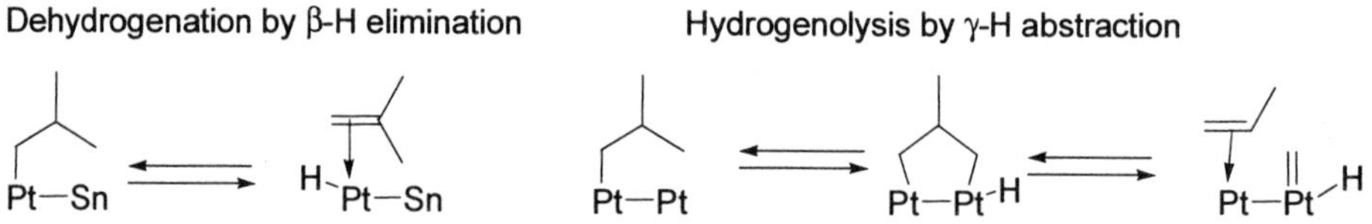

Dehydrogenation by β-H elimination   Hydrogenolysis by γ-H abstraction

SCHEME 21.9

As a result, a precise description of the active site, including site isolation, allows one to build a structure-activity relationship and to improve the catalytic performances of heterogeneous catalysts.

These examples illustrate that it is possible to tune the catalyst's selectively and lifetime by selective poisoning of the metal surface.

21.4.3. Application of Surface Organometallic Chemistry on Metals to Depollution

Based on these concepts, The RAM II (removal of arsine and mercury) process was developed by IFP for the simultaneous removal of arsenic, mercury and lead from contaminated feedstocks upstream of aromatics or steam-cracking units.[80, 81] For this purpose, the contaminated hydrocarbon is passed under an hydrogen pressure successively through two reactors filled by sulfided nickel based solid (Scheme 21.10, next page).

The temperature of the first bed is about 170 °C. The organometallic compounds are decomposed into elemental As and Hg. Arsenic is trapped in the bulk of the solid as NiAs (nickeline; DRX) and Hg is vaporized. The temperature of the second bed is lower (about 80 °C) and Hg is fixed as HgS (cinnabar). Using this system less than 3 ppb of each contaminant remain in the process effluent.

The mechanism of de-metallation has been fully studied and understood : triphenylarsine and diphenylmercury are converted on Ni under H$_2$ into organic by-products and As$^{(0)}$ or Hg$^{(0)}$ (Scheme 21.11, next page). It corresponds to the mechanism described previously for the reaction of organotin agents and metal surfaces (see Scheme 21.8 and discussion).[81]

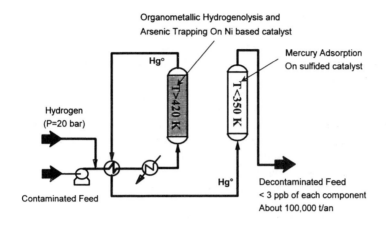

SCHEME 21.10

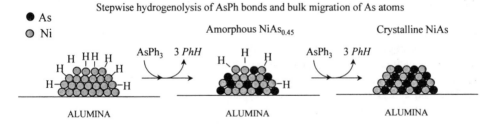

SCHEME 21.11. Mechanism of de-arsenification of petroleum feeds.

21.5. SUMMARY AND OUTLOOK

We have shown that the SOMC approach allows one to improve activity, selectivity and life time in a variety of reactions such as olefin metathesis, olefin asymmetric epoxidations and olefin polymerisation etc... For example, in the area of olefin metathesis, controlling the formation of the "active" site at the nanoscale level led to a well-defined silica supported alkylidene Re-complex [(≡SiO)Re(≡CtBu)(=CHtBu)(CH$_2$tBu)], which is typically more active, selective and functional group-tolerant than the existing heterogeneous and homogeneous Re-based catalysts.[18, 30] Since the coordination sphere of the precursor is well-defined, it displays a clean and identified initiation step (cross-metathesis), and should give a chance to investigate the origin of its reactivity, its mode of deactivation and a way for its regeneration.

Furthermore, *Surface Organometallic Chemistry* (SOMC) can generate entities that display reactivities never observed either in heterogeneous or in homogeneous catalysis. For instance, metal hydrides, which correspond in fact to isolated moieties, are stabilized only by the surface, which acts as a ligand. Therefore these species are very electrophilic and readily react with the inert bonds of alkanes, leading to depolymerisation[59] or alkane metathesis catalysts.[62] These two novel reactions have therefore been discovered through the mastering of the metal grafting process on a surface via a nano-scopic scale-control of the relevant parameters (reactivity at the grafting site, coordination sphere of the bound moiety), commonly called SOMC.

Additionally, SOMC allows one to control the properties of porous materials such as zeolithes or well-defined mesoporous materials by selective derivatisation of the inner or outer surface of such materials. This concept has been illustrated by the generation of separation materials for alkane mixtures through controlling the pore size or the preparation of shape selective catalysts, which require a passivated outer surface.[64]

Finally, in the case of metallic surfaces, it is now possible to control the local environment of the active sites, and therefore tailor-make a catalyst – at the nanoscale level – to give it the desired selectivity.[75]

Nonetheless, while this chemistry shows promise, it is still young and many challenges must be met to further transpose some of these systems to industrial processes. Moreover, there are plenty of current industrial processes that will require major changes to meet the criterion of sustainable technology, such as high selectivity, high space time yield, environmentally friendliness, and low energy cost. This will require either a dramatic change of processes and therefore the discovery of new systems or major improvements of actual processes. What is certain is that innovation in catalysis, a major player in sustainable technology, is crucially needed. Since catalysis is a molecular phenomenon, it seems that designing catalysts at a molecular level must help elaborating and improving processes, which must be highly selective, environmentally friendliness, and cost effective. Heterogeneous catalysis has a key role to play, and therefore it will be necessary to generate and understand these types of catalysts with an atomic precision directed towards process application. That is, building a process from the atom that constitutes the active site, to the designed material that will be used in actual processes.

This challenge will require a tremendous effort, spanning science from its fundamentals to the very applied aspects, that is from how to make and understand these new types of heterogeneous catalysts on one side {molecular, supramolecular, solid state chemistry and organometallic chemistry combined with spectroscopies (*in situ*) and molecular modelling} to prepare them in a large scale to use them in pilot plants (chemical kinetics, and reactor design). Obviously it will require a multidisciplinary approach, which should bring experts from the various fields of science to engage in this challenge.

ACKNOWLEDGMENTS

Past and present research associates of Laboratoire de Chimie Organométallique de Surface are gratefully thanked. Financial support from CNRS, ESCPE-Lyon, Region Rhône-Alpes and industrial partners is gratefully acknowledged.

REFERENCES

1. D.G.H. Ballard, Pi and sigma transition metal carbon compounds as catalysts for the polymerization of vinyl monomers and olefins, *Advances in Catalysis* **23**, 263-325 (1973).
2. J.P. Candlin and H. Thomas, Supported organometallic catalysts, *Advances in Chemistry Series* **132** (Homogeneous Catal.-2, Symp., 1973), 212-239 (1974).
3. Y.I. Yermakov, B.N. Kuznetsov, and V.A. Zakharov, *Studies in Surface Science and Catalysis* **8** (Catalysis by Supported Complexes), 522 (1981).
4. Y. Iwasawa, Tailored Metal Catalysis p. 333 (1986).
5. J. Evans, Reaction of organometallics with surfaces of metal oxides, *NATO ASI Series, Series C: Mathematical and Physical Sciences*, **231** (Surf. Organomet. Chem.: Mol. Approaches Surf. Catal.), 47-73 (1988).
6. For a recent review see: C. Copéret, M. Chabanas, R.P. Saint-Arroman, and J.-M. Basset, Homogeneous and heterogeneous catalysis: Bridging the gap through surface organometallic chemistry, *Angewandte Chemie, International Edition* **42** (2), 156-181 (2003).
7. G. Ertl, H. Knoezinger, *Handbook of Heterogeneous Catalysis* Vol. 5, p. 2800 (1997).
8. M.E. Bartram, T.A. Michalske, and J.W. Rogers, Jr., A reexamination of the chemisorption of trimethylaluminum on silica, *Journal of Physical Chemistry* **95** (11), 4453-4463 (1991).
9. B.A. Morrow, Surface groups on oxides, *Studies in Surface Science and Catalysis* **57** (A) (Spectrosc. Charact. Heterog. Catal., Part A), 161-224 (1990).
10. E.F. Vansant, P. Van Der Voort and K.C. Vrancken, *Studies in Surface Science and Catalysis*, **93** (Characterization and Chemical Modification of the Silica Surface), 572 (1995).
11. B.A. Morrow and I.D. Gay, Infrared and NMR characterization of the silica surface, *Surfactant Science Series*, **90** (Adsorption on Silica Surfaces), 9-33 (2000).
12. V. Dufaud, G.P. Niccolai, J. Thivolle-Cazat, and J.-M. Basset, Surface Organometallic Chemistry of Inorganic Oxides: The Synthesis and Characterization of ($\equiv$SiO)Ta(=CHC(CH$_3$)$_3$)(CH$_2$C(CH$_3$)$_3$)$_2$ and ($\equiv$SiO)$_2$Ta(=CHC(CH$_3$)$_3$)(CH$_2$C(CH$_3$)$_3$), *Journal of the American Chemical Society*, **117** (15), 4288-4294 (1995).
13. L. Lefort, M. Chabanas, O. Maury, D. Meunier, C. Copéret, J. Thivolle-Cazat, and J.-M. Basset, Versatility of silica used as a ligand: effect of thermal treatments of silica on the nature of silica-supported alkyl tantalum species, *Journal of Organometallic Chemistry* **593-594**, 96-100 (2000).
14. M. Chabanas, E.A. Quadrelli, B. Fenet, C. Copéret, J. Thivolle-Cazat, J.-M. Basset, A. Lesage, and L. Emsley, Molecular insight into surface organometallic chemistry through the combined use of 2D HETCOR solid-state NMR spectroscopy and silsesquioxane analogs, *Angewandte Chemie, International Edition*, **40** (23), 4493-4496 (2001).
15. M. Chabanas, D. Alcor, E. Gautier, C. Copéret, J.M. Basset, A. Lesage, S. Hediger, L. Emsley, and W. Lukens, unpublished results.
16. R. Petroff Saint-Arroman, M. Chabanas, A. Baudouin, C. Copéret, J.-M. Basset, A. Lesage, and L. Emsley, Characterization of Surface Organometallic Complexes Using High Resolution 2D Solid-State NMR Spectroscopy. Application to the Full Characterization of a Silica Supported Metal Carbyne: $\equiv$SiO-Mo(=C-Bu-t)(CH$_2$-Bu-t)$_2$, *Journal of the American Chemical Society*, **123** (16), 3820-3821 (2001).
17. M. Chabanas, Ph.D. Thesis, Université Claude Bernard Lyon I, 2001.
18. M. Chabanas, A. Baudouin, C. Copéret, and J.-M. Basset, A Highly Active Well-Defined Rhenium Heterogeneous Catalyst for Olefin Metathesis Prepared via Surface Organometallic Chemistry, *Journal of the American Chemical Society* **123** (9), 2062-2063 (2001).
19. A. Lesage, L. Emsley, M. Chabanas, C. Copéret, and J.-M. Basset, Observation of a H-agostic bond in a highly active rhenium-alkylidene olefin metathesis heterogeneous catalyst by two-dimensional solid-state NMR spectroscopy, *Angewandte Chemie, International Edition*, **41** (23), 4535-4538 (2002).
20. M. Chabanas, A. Baudouin, C. Copéret, J.-M. Basset, W. Lukens, A. Lesage, S. Hediger, and L. Emsley, Perhydrocarbyl ReVII Complexes: Comparison of Molecular and Surface Complexes, *Journal of the American Chemical Society* **125** (2), 492-504 (2003).
21. F.J. Feher and T.A. Budzichowski, Silasesquioxanes as ligands in inorganic and organometallic chemistry, *Polyhedron* **14** (22), 3239-3253 (1995).
22. P.T. Wolczanski, Chemistry of electrophilic metal centers coordinated by silox (tBu$_3$SiO), tritox (tBu$_3$CO) and related difunctional ligands, *Polyhedron* **14** (22), 3335-3362 (1995).
23. P.D. Lickiss, The synthesis and structure of organosilanols, *Advances in Inorganic Chemistry* **42**, 147-262 (1995).

24. R. Murugavel, A. Voigt, M.G. Walawalkar, and H.W. Roesky, Hetero- and Metallasiloxanes Derived from Silanediols, Disilanols, Silanetriols, and Trisilanols, *Chemical Reviews (Washington, D. C.)*, **96** (6), 2205-2236 (1996).

25. P.G. Harrison, Silicate cages: precursors to new materials, *Journal of Organometallic Chemistry* **542** (2), 141-183 (1997).

26. L. King and A.C. Sullivan, Main group and transition metal compounds with silanediolate $[R_2SiO_2]^{2-}$ and .alpha.,.omega.-siloxane diolate $[O(R_2SiO)n]^{2-}$ ligands, *Coordination Chemistry Reviews* **189**, 19-57 (1999).

27. V. Lorenz, A. Fischer, S. Giessmann, J.W. Gilje, Y. Gun'ko, K. Jacob, and F.T. Edelmann, Disiloxanediolates and polyhedral metallasilsesquioxanes of the early transition metals and f-elements, *Coordination Chemistry Reviews* **206-207**, 321-368 (2000).

28. H.C.L. Abbenhuis, Advances in homogeneous and heterogeneous catalysis with metal-containing silsesquioxanes, *Chemistry--A European Journal* **6** (1), 25-32 (2000).

29. B. Marciniec and H. Maciejewski, Transition metal-siloxide complexes; synthesis, structure and application to catalysis, *Coordination Chemistry Reviews* **223**, 301-335 (2001).

30. M. Chabanas, C. Copéret, and J.-M. Basset, Re-based heterogeneous catalysts for olefin metathesis prepared by surface organometallic chemistry: Reactivity and selectivity, *Chemistry--A European Journal* **9** (4), 971-975 (2003).

31. The activity of the rhenium oxides supported on silica is usually poor under 200°C. For some examples see this reference and next four ones: A.W. Aldag, C.J. Lin, and A. Clark, On the number of active sites for the disproportionation of ethylene on supported rhenium oxide catalysts. *Recueil des Travaux Chimiques des Pays-Bas* **96** (11), 27-30 (1977).

32. N. Tsuda and A. Fujimori, Metathesis of propylene over unsupported rhenium trioxide, *Journal of Catalysis* **69** (2), 410-417 (1981).

33. L.G. Duquette, R.C. Cieslinski, C.W. Jung, and P.E. Garrou, ESCA studies on silica- and alumina-supported rhenium oxide catalysts, *Journal of Catalysis* **90** (2), 362-365 (1984).

34. P.S. Kirlin and B.C. Gates, A stable supported rhenium metathesis catalyst derived from tetrakis(tricarbonylhydroxyrhenium) on silica, *Journal of the Chemical Society, Chemical Communications* (5), 277-279 (1985).

35. R. Edreva-Kardzhieva and A. Andreev, Formation of the active form of rhenium heptoxide/silica catalyst in olefin metathesis, *Journal of Catalysis* **97** (2), 321-329 (1986).

36. For a comparison of Re_2O_7 supported on alumina, see: Y. Chauvin and D. Commereuc, Chemical counting and characterization of the active sites in the rhenium oxide/alumina metathesis catalyst, *Journal of the Chemical Society, Chemical Communications* (6), 462-464 (1992).

37. J.C. Mol, Olefin metathesis over supported rhenium oxide catalysts, *Catalysis Today* **51** (2), 289-299 (1999).

38. R. Toreki, G.A. Vaughan, R.R. Schrock, and W.M. Davis, Metathetical reactions of rhenium(VII) alkylidene-alkylidyne complexes of the type $Re(CR')(CHR')[OCMe(CF_3)_2]_2$ (R' = CMe$_3$ or CMe$_2$Ph) with terminal and internal olefins, *Journal of the American Chemical Society* **115** (1), 127-137 (1993).

39. A.M. LaPointe and R.R. Schrock, Alkyl, Alkylidene, and Alkylidyne Complexes of Rhenium, *Organometallics* **14** (4), 1875-1884 (1995).

40. B.T. Flatt, R.H. Grubbs, R.L. Blanski, J.C. Calabrese, and J. Feldman, Synthesis, Structure, and Reactivity of a Rhenium Oxo-Vinylalkylidene Complex. *Organometallics* **13** (7), 2728-2732 (1994).

41. D. Commereuc, New homogeneous rhenium-based metathesis catalysis as models of the rhenium on alumina heterogeneous catalyst, *Journal of the Chemical Society, Chemical Communications* (7), 791-792 (1995).

42. A. Andresen, H.G. Cordes, J. Herwig, W. Kaminsky, A. Merck, R. Mottweiler, J. Pein, H. Sinn, and H.J. Vollmer, Halogen-free soluble Ziegler catalysts for ethylene polymerization. Control of molecular weight by the choice of the reaction temperature, *Angewandte Chemie* **88** (20), 689-690 (1976).

43. For a review see: W. Kaminsky and Editor, Metalorganic Catalysts for Synthesis and Polymerisation. (Proceedings of an International Symposium held 13-17 September 1998, in Hamburg, Germany.). p. 674 (1999).

44. M. Jezequel, V. Dufaud, M.J. Ruiz-Garcia, F. Carrillo-Hermosilla, U. Neugebauer, G.P. Niccolai, F. Lefebvre, F. Bayard, J. Corker, S. Fiddy, J. Evans, J.-P. Broyer, J. Malinge, and J.-M. Basset, Supported Metallocene Catalysts by Surface Organometallic Chemistry. Synthesis, Characterization, and Reactivity in Ethylene Polymerization of Oxide-Supported Mono- and Biscyclopentadienyl Zirconium Alkyl Complexes: Establishment of Structure/Reactivity Relationships, *Journal of the American Chemical Society* **123** (15), 3520-3540 (2001).

45. J.F. Walzer, Jr., Supported metallocene catalyst composition, in *U.S. patent* No. 285,380 (1997) (Exxon Chemical Patents Inc., USA), (abandoned)

46. S.J. Lancaster, S.M. O'Hara, and M. Bochmann, Heterogenised MAO-free Metallocene Catalysts, in *Metalorganic Catalysts for Synthesis and Polymerisation*, Edited by Kaminski, (Springer, 199) p. 413.

47. M. Bochmann, G.J. Pindado, and S.J. Lancaster, The versatile chemistry of metallocene polymerisation catalysts: new developments in half-sandwich complexes and catalyst heterogenisation, *Journal of Molecular Catalysis A: Chemical* **146** (1-2), 179-190 (1999).

48. N. Millot, A. Cox, C.C. Santini, Y. Molard, and J.-M. Basset, Surface organometallic chemistry of main group elements: selective synthesis of silica supported [≡Si-OB(C₆F₅)₃]⁻[HNEt₂Ph]⁺, *Chemistry--A European Journal* **8** (6), 1438-1442 (2002).

49. D. Meunier, A. Piechaczyk, A. De Mallmann, and J.-M. Basset, Silica-supported tantalum catalysts for asymmetric epoxidations of allyl alcohols, *Angewandte Chemie, International Edition* **38** (23), 3540-3542 (1999).

50. V. Vidal, A. Theolier, J. Thivolle-Cazat, J.-M. Basset, and J. Corker, Synthesis, Characterization, and Reactivity, in the C-H Bond Activation of Cycloalkanes, of a Silica-Supported Tantalum(III) Monohydride Complex: (≡SiO)₂Taᴵᴵᴵ-H, *Journal of the American Chemical Society* **118** (19), 4595-4602 (1996).

51. J. Corker, F. Lefebvre, C. Lecuyer, V. Dufaud, F. Quignard, A. Choplin, J. Evans, and J.-M. Basset, Catalytic cleavage of the C-H and C-C bonds of alkanes by surface organometallic chemistry: an EXAFS and IR characterization of a Zr-H catalyst, *Science (Washington, D. C.)* **271** (5251), 966-969 (1996).

52. C. Rosier, G.P. Niccolai, and J.-M. Basset, Catalytic Hydrogenolysis and Isomerization of Light Alkanes over the Silica-Supported Titanium Hydride Complex (≡SiO)₃TiH, *Journal of the American Chemical Society* **119** (50), 12408-12409 (1997).

53. S.A. Holmes, F. Quignard, A. Choplin, R. Teissier, and J. Kervennal, Tetraneopentyltitanium derived supported catalysts. Part 1, Synthesis and catalytic properties for the epoxidation of cyclohexene with aqueous hydrogen peroxide, *Journal of Catalysis* **176** (1), 173-181 (1998).

54. L. d' Ornelas, S. Reyes, F. Quignard, A. Choplin, and J.M. Basset, Hafnium-hydride complexes anchored to silica: catalysts for low-temperature hydrogenolysis of alkanes and hydrogenation of olefins, *Chemistry Letters* (11), 1931-1934 (1993).

55. G.P. Niccolai and J.-M. Basset, Primary selectivity in the activation of the carbon-hydrogen bonds of propane by silica-supported zirconium hydride, *Applied Catalysis, A: General* **146** (1), 145-156 (1996).

56. G.L. Casty, M.G. Matturro, G.R. Myers, R.P. Reynolds, and R.B. Hall, Hydrogen/Deuterium Exchange Kinetics by a Silica-Supported Zirconium Hydride Catalyst: Evidence for a .sigma.-Bond Metathesis Mechanism, *Organometallics* **20** (11), 2246-2249 (2001).

57. L. Lefort, C. Copéret, M. Taoufik, J. Thivolle-Cazat, and J.-M. Basset, H/D exchange between CH4 and CD4 catalysed by a silica supported tantalum hydride, (≡SiO)₂Ta-H, *Chemical Communications (Cambridge)* (8), 663-664 (2000).

58. C. Lecuyer, F. Quignard, A. Choplin, D. Olivier, and J.M. Basset, Organometallic chemistry on oxide surfaces. Selective, catalytic low-temperature hydrogenolysis of alkanes by a highly electrophilic, silica-gel-supported zirconium hydride, *Angewandte Chemie* **103** (12), 1692-1694 (1991) (See also Angew. Chem., Int. Ed. Engl. **103** (12), 1660-1691 (1991).

59. V. Dufaud and J.-M. Basset, Catalytic hydrogenolysis at low temperature and pressure of polyethylene and polypropylene to diesels or lower alkanes by a zirconium hydride supported on silica-alumina: a step toward polyolefin degradation by the microscopic reverse of Ziegler-Natta polymerization, *Angewandte Chemie, International Edition* **37** (6), 806-810 (1998).

60. M. Chabanas, V. Vidal, C. Copéret, J. Thivolle-Cazat, and J.-M. Basset, Low-temperature hydrogenolysis of alkanes catalyzed by a silica-supported tantalum hydride complex, and evidence for a mechanistic switch from Group IV to Group V metal surface hydride complexes, *Angewandte Chemie, International Edition* **39** (11), 1962-1965 (2000).

61. C. Copéret, O. Maury, J. Thivolle-Cazat, and J.-M. Basset, Sigma.-Bond metathesis of alkanes on a silica-supported tantalum(V) alkyl alkylidene complex: first evidence for alkane cross-metathesis, *Angewandte Chemie, International Edition* **40** (12), 2331-2334 (2001).

62. V. Vidal, A. Theolier, J. Thivolle-Cazat, and J.-M. Basset, Metathesis of alkanes catalyzed by silica-supported transition metal hydrides, *Science (Washington, D. C.)* **276** (5309), 99-102 (1997).

63. O. Maury, L. Lefort, V. Vidal, J. Thivolle-Cazat, and J.-M. Basset, Metathesis of alkanes: evidence for degenerate metathesis of ethane over a silica-supported tantalum hydride prepared by surface organometallic chemistry, *Angewandte Chemie, International Edition* **38** (13/14), 1952-1955 (1999).

64. F. Lefebvre, A. De Mallmann, and J.-M. Basset, Modification of the adsorption and catalytic properties of molecular sieves by reaction with organometallic complexes, *European Journal of Inorganic Chemistry*, (3), 361-371 (1999).

65. X. Wang, Ph.D. Thesis, Univerité Claude Bernard Lyon I, 1999.

66. C. Nédez, Ph.D. Thesis, Université Claude Bernard Lyon I, 1992.

67. C. Nédez, A. Theolier, F. Lefebvre, A. Choplin, J.M. Basset, J.F. Joly, and E. Benazzi, Chemical grafting of tin alkyl complexes on the external surface of mordenite: A method for controlling the size of the pore entrances of zeolites, *Microporous Materials* **2** (4), 251-259 (1994).

68. E. Benazzi, S. De Tavernier, P. Beccat, J.F. Joly, C. Nédez, A. Choplin, and J.M. Basset, Selective isomerization to xylenes, *Chemtech* **24** (10), 13-18 (1994).

69. E. Benazzi, C. Travers, A. Choplin, J.F. Joly, and J.M. Basset, Omega zeolite catalysts containing Group IIA, IVB, IIB or IVA metals for the isomerization of C8 aromatic fractions, *Eur. Pat.* 569268 (1993) (Institut Francais du Petrole).

70. M. Adachi, F. Lefebvre, C. Schott-Darie, H. Kessler, and J.M. Basset, Modification of hydrocarbon sorption properties of cloverite by grafting organotin complexes, *Applied Surface Science* **121** (122), 355-359 (1997).

71. M. Adachi, J. Corker, H. Kessler, F. Lefebvre, and J.M. Basset, Study of cloverite molecular sieves modified by grafting organometallic complexes inside the pores. I. Synthesis and characterization of cloverite modified by grafting -SnR$_3$ fragments (R = Me, Et, *n*-Bu, Cy), *Microporous and Mesoporous Materials* **21** (1-3), 81-90 (1998).

72. M. Adachi, J. Corker, H. Kessler, A. De Mallmann, F. Lefebvre, and J.M. Basset, Study of cloverite molecular sieves modified by grafting organometallic complexes inside the pores II. Adsorption properties of cloverite modified by grafting -SnR$_3$ fragments (R = Me, Et, *n*-Bu, Cy), *Microporous and Mesoporous Materials* **28** (1), 155-162 (1999).

73. X. Wang, H. Zhao, F. Lefebvre, and J.-M. Basset, Surface organometallic chemistry of tin: grafting reaction of Sn(CH$_3$)$_4$ in HY zeolite supercage, *Chemistry Letters* (10), 1164-1165 (2000).

74. P. Lesage, O. Clause, P. Moral, B. Didillon, J.P. Candy, and J.M. Basset, Surface organometallic chemistry on metals: preparation of bimetallic catalyst by controlled hydrogenolysis of Sn(*n*-C$_4$H$_9$)$_4$ on a Ni/SiO$_2$ catalyst, *Journal of Catalysis* **155** (2), 238-248 (1995).

75. B. Didillon, C. Houtman, T. Shay, J.P. Candy, and J.M. Basset, Surface organometallic chemistry on metals. Evidence for a new surface organometallic material, Rh[Sn(*n*-C$_4$H$_9$)$_x$]$_y$/SiO$_2$, obtained by controlled hydrogenolysis of tetra-*n*-butylstannane on a rhodium/silica catalyst, *Journal of the American Chemical Society* **115** (21), 9380-9388 (1993).

76. F. Humblot, D. Didillon, F. Lepeltier, J.P. Candy, J. Corker, O. Clause, F. Bayard, and J.M. Basset, Surface Organometallic Chemistry on Metals: Formation of a Stable .ident.Sn(*n*-C$_4$H$_9$) Fragment as a Precursor of Surface Alloy Obtained by Stepwise Hydrogenolysis of Sn(*n*-C$_4$H$_9$)$_4$ on a Platinum Particle Supported on Silica, *Journal of the American Chemical Society* **120** (1), 137-146 (1998).

77. G. Meitzner, G.H. Via, F.W. Lytle, S.C. Fung, and J.H. Sinfelt, Extended x-ray absorption fine structure (EXAFS) studies of platinum-tin catalysts, *Journal of Physical Chemistry* **92** (10), 2925-2932 (1988).

78. B. Didillon, A. El Mansour, J.P. Candy, J.P. Bournonville, and J.M. Basset, Surface organometallic chemistry on metals: selective hydrogenation of citral to geraniol and nerol on tin-modified silica-supported rhodium, *Studies in Surface Science and Catalysis* **59** (Heterog. Catal. Fine Chem. 2), 137-143 (1991).

79. A. El Mansour, J.P. Candy, J.P. Bournonville, O.A. Ferretti, and J.M. Basset, Selective hydrogenation of esters to alcohols with a rhodium(III) oxide, Sn(*n*-C$_4$H$_9$)$_4$, and SiO$_2$ supported catalyst containing isolated active sites, *Angewandte Chemie* **101** (3), 360-362 (1989).

80. Y.A. Ryndin, J.P. Candy, B. Didillon, L. Savary, and J.M. Basset, Surface Organometallic Chemistry on Metals Applied to the Environment: Hydrogenolysis of AsPh$_3$ with Nickel Supported on Alumina, *Journal of Catalysis* **198** (1), 103-108 (2001).

81. P. Sarrazin, C.J. Cameron, Y. Barthel, and M.E. Morrison, Processes prevent detrimental effects from arsenic and mercury in feedstocks, *Oil & Gas Journal* **91** (4), 86-90 (1993).

22

Modelling Transition Metal Nanoparticles: the Role of Size Reduction in Electronic Structure and Catalysis

L. Guczi, Z. Pászti and G. Pető[*]

22.1. INTRODUCTION

Mono- or multiphase polycrystalline solid materials whose size falls – at least in one dimension – into the nanometer range (typically 1–100 nm) are regarded as representatives of different kinds of nanomaterials. Depending on the number of dimensions in the nanometer domain we can distinguish (i) nanoparticles, (ii) fibrillar structures e.g. nanotubes and (iii) layered structures.[1, 2]

In order to reach the region of nanoparticles the size of the domain should decrease. During size reduction the surface/volume ratio is enlarged and an interface is formed which is transformed into a metastable system through a series of quasi equilibrium states. The key issue in the field of nanoparticles is the easy functionalization of the interface due to its high excess free energy acquired. The surface irregularities, i.e. the presence of steps, kinks, terraces, as well as the "dangling" bonds of the atoms located at these sites, make it possible to create highly reactive species. The energy introduced into the system for modification of the surface may originate from various sources. For example, by introducing physical or chemical energies to the system active surface sites e.g. catalysts can be prepared.[3-5]

*László Guczi and Zoltán Pászti, Department of Surface Chemistry and Catalysis, Institute of Isotope and Surface Chemistry, Chemical Research Centre, Hungarian Academy of Sciences, P. O. Box 77, H-1525 Budapest, Hungary, Gábor Pető, Institute of Technical Physics and Materials Science, Hungarian Academy of Sciences, P. O. Box 49, H-1525 Budapest, Hungary

Size reduction of metal particles results in several changes of the surface properties. The primary change is observed in the electronic properties of the metal particles characterized by the *shift in the characteristic structures of electron spectra* measured by UPS, XPS and Auger-electron spectroscopy. Furthermore, morphology of the metal nanoparticles is highly sensitive to the environment, such as *ion-metal interaction* (e.g. metal/support interaction) influencing also the electronic properties of the metal surface. Since metal nanoparticles have a short-range ordering, these particles are in metastable state, thus they must be *stabilized against coalescence* to prevent formation of large metal particles.

As any materials interact with their environment through solid/gas, solid/liquid, and solid/solid interfaces, the nanometer scale surface created can be modified to perform certain functions. The modifications are usually only effective in the few nanometer deep surface layers.

This Chapter highlights the development of new model nano-structured materials with functionalized interfaces to promote highly efficient and specific catalysts. The functional characteristics of nanoparticles (morphology, electronic structure) are thoroughly examined at atomic level by means of various techniques, e.g. TEM, UPS, XPS, AES, AFM, etc. indicating a change in the metal nanoparticles/support interface which has also a consequence for the catalytic activity. In addition to physical methods, correlation is sought between physical characteristics and catalytic reaction monitoring by the turnover frequency (TOF) to elucidate the role of surface ordering, restructuring, etc. The major factors controlling the catalytic activity are the size of metal/support interface. Moreover, size dependent changes in the electronic structure of the metal nanoparticles are also expected to strongly influence their catalytic properties. Along with size reduction the metal d-valence band structure is investigated and the actual size is determined by ultraviolet photoelectron spectroscopy (UPS) and TEM, respectively. The chemisorption and catalytic properties in various systems are also determined.

22.2. GENERAL PRINCIPLES

Application of small metal particles has attracted the attention of scientists for a long time. In the seventies Turkevich prepared monodispersed gold particles,[6] and later, using molecular transition metal carbonyl clusters,[7] the importance of small nanoparticles increased considerably. Advantages of small metal particles are (i) short range ordering, (ii) enhanced interaction with environments due to the high number of dangling bonds (iii) great variety of the valence band electron structure and (iv) self-structuring for optimum performance in chemisorption and catalysis. One of the crucial points is whether turnover frequency measured for a given catalytic reaction increases or decreases as the particle size is diminished.

In the literature, there are several attempts to measure this relationship. Che and Bennett found that e.g. for ethylene hydrogenation over palladium catalyst a maximum in TOF was shown as a function of particle size, the maximum being at a particle of 0.6 nm diameter.[8] The Japanese school[9-11] observed controversial data on several systems[10] at which TOF increased with reducing the particle size, although opposite results were also measured.[12]

There are two fundamental problems with nanoparticles and all conflicting data originated from this. First, in most cases the size of the small metal particles is not uniform in the sampie and thus the TOF is calculated from an average particle size. Consequently, the rate from which the TOF is calculated refers to an average rate, the composition of which is unknown. Second, the spacing between the particles is also non-uniform. This is why the compiled data collected by Ribeiro et al.[13] are randomly dispersed. Accordingly, in the ideal case both the size and the spacing of metal particles on a support should be uniform and regular.

Now, the question arisen is how the size of metal particles can be reduced to the size-range of nanoparticles. There are several methods available from the literature, however, these are not the ones, which – in most cases – result in uniform size and spacing on the surface. Nevertheless, it is still worth mentioning (i) the zeolite encaged metal clusters,[14-17] (ii) the organometallic molecular clusters deposited on inorganic oxide supports,[18-20] (iii) the inverse micelle technique.[21, 22] Recent application of thin film technology based techniques offers a new approach to fabricate uniform size of metal nanoparticles as detailed and discussed later in this Chapter.[23-27]

When metal nanoparticles are inserted into zeolite supercages, the size of the metal particles is confined according to the size of the supercage. However, after reduction of the precursor metal ions in a stream of hydrogen, the protons replacing the metal ions in the cation exchange positions interfere with the metal particles, thereby influencing their chemisorption and catalytic properties.

As an alternative approach towards the above requirement, Somorjai introduced the method of electron lithography,[28] which represents the most advanced so called "HIGHTECH" sample preparation technique. The method ensures uniform particle size and spacing: e.g. Pt particles of 25 nm size could be placed with 50 nm separation. This array showed a uniform activity similar to those measured on single crystal in ethylene hydrogenation. The only difficulty with the method is that the particle size is, so far, not small enough.[10, 29, 30]

At the same time, thin film technology offers a very interesting way for reaching the nanometric size regime by *removing* atoms from already existing, not necessarily nanosized particles. Although the possibilities and limitations of this strategy are hardly known, it could be technologically very important. The main difficulty lies in how to choose the source of energy, which initiates material removal from the original particles. Nevertheless, recent studies indicate that irradiation of nanoparticles with intense laser beams can induce significant reduction in the particle size,[40-42] proving the feasibility of this approach.

Due to its sputtering effect, *ion bombardment* is another potential material removing tool. Because of the inevitable ballistic and chemically guided mixing processes, it was conventionally regarded as a predominantly destructive process. However, successful ion beam induced nanoparticle preparation experiments carried out by using high energy bombarding beams[43-49] suggest that sputtering with low energy ion beams may similarly offer a valuable method for post-deposition modification of the properties of already existing surface nanosystems. Unfortunately, in spite of its conceptual simplicity, there are only very few indications of the feasibility of this nanoparticle preparation approach.[26, 27, 50-54]

Generally, if metal loading of a catalyst is diminished, the metal particle size decreases. As a consequence of the size reduction, (i) there is a shift in the valence bands or (ii) the valence band structure is altered due to the change in the bond energy of electrons in the d-band. The change in the valence band electronic structure may mean that partial electron deficiency is built up on the metal particles, which influences the mode of chemisorption of the substrates. For instance, CO chemisorption on metals could be dissociative or associative controlled by the local density of states on the metal (LDOS). The carbon-metal bond strength can be altered as shown in Scheme 22.1. According to the Blyholder model the CO molecule donates 5σ–electrons to the metal empty d-orbital and the metal back donates d-electrons to the $2\pi^*$ orbital.

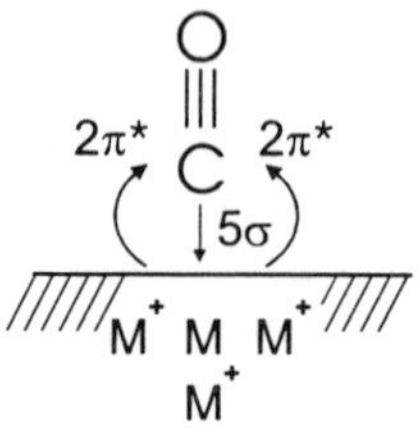

SCHEME 22.1.

22.3. PREPARATION OF SUPPORTED NANOPARTICLES BY PULSED LASER DEPOSITION AND ION ETCHING

In the followings, we will give a detailed example for nanoparticle preparation by pulsed laser deposition as well as by low energy ion etching of an island thin film.

22.3.1. Nanoparticle Formation during Pulsed Laser Deposition[24, 25]

The basic process of pulsed laser deposition is presented in Fig. 22.1. The principle of the method is to let a nanosecond range laser pulse interact with the target to be used as material for the nanoparticles. The essential part of the technique is a Nd:glass laser with 2 J pulse energy and a 30 nanosecond pulse length. By hitting the surface of the target the following events occur during the 30 nsec pulse. First, energy in the pulse is transmitted and absorbed by the target surface, then the outermost layer is melted followed by vaporization and finally the plasma ejection toward the substrate which is located at various angles with respect to the direction of the beam. The increase in the surface temperature of the target metal depends on the laser beam penetration, thermal diffusion and the rate of energy deposition on the target (laser pulse width).

Although pulsed laser deposition with nanosecond pulses is still a thermal process, it differs considerably from the conventional (resistively or electron beam induced) evaporation. The local temperature at the target surface can be very high, around 10000 K, which allows evaporation and stoichiometric deposition of virtually all kind of materials. Moreover, the expanding cloud of the evaporated material, known as plume, may interact with the laser beam, and even the neutral atoms can acquire kinetic energies in the order of 1–10 eV. The kinetic energy of the ionized species is in the range of 100–

1000 eV.[55] When these energetic species start to condense on a substrate, their high mobility allows them to form very compact, high structural quality films.

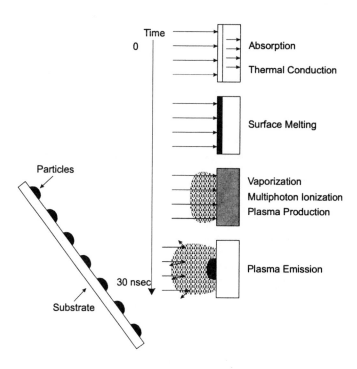

FIGURE 22.1. Technical arrangement for fabrication of nanoparticles by Pulsed Laser deposition (PLD). Reprinted with permission from Ref.[23]. Copyright (1997) American Chemical Society.

The kinetic energy of the evaporated species can be most easily influenced by the presence of an inert gas, which reduces their mean free path from kilometers (in high vacuum) to microns (in the 1–10 mbar range). Consequently, pressure is definitely one of the crucial parameters regulating the events after the laser pulse, and the structure of the deposited metal layers. The effect of the background pressure on the properties of pulsed laser deposited layers was studied in an oil-free evaporation chamber with 3×10^{-7} mbar base pressure, equipped with a quartz oscillator deposition monitor. Amorphous carbon covered micro-grids or pre-thinned silicon pieces were applied as substrates suitable for transmission electron microscopy studies.

In Fig. 22.2 we present the results of copper evaporation while changing the ambient Ar pressure between 0–10 mbar. On the vacuum deposited sample (Fig. 22.2 (a)) more or less regular, compact islands can be seen, as one would expect in the case of an usual thin film growth process involving highly energetic species. This indicates, that pulsed laser deposition in vacuum alone is capable for preparation of surface nanostructures if the amount of the deposited material remains low.[54] In 1 mbar (Fig. 22.2 (b)), instead of the islands, several lonely particles with sizes at about 5 nm and a few aggregates containing not more than 20–30 particles with diameters around 10 nm can be observed.

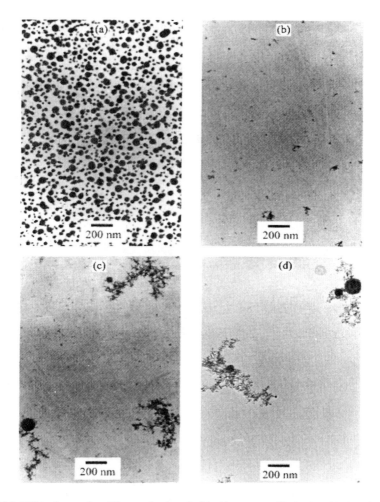

FIGURE 22.2. TEM micrographs of Cu samples deposited in (a): vacuum, (b): 1 mbar, (c): 2.5 mbar and (d): 10 mbar Ar.

At 2.5 mbar (Fig. 22.2 (c)) a strongly different morphology was observed, which remained unchanged up to 10 mbar (Fig. 22.2 (d)). The dominating structural elements are huge aggregates containing lots of nanoparticles with sizes around 4–6 nm, sometimes also including several large particles (up to 100–200 nm).

As a next example, we can use silver as target material instead of copper. The morphology of the vacuum deposited Ag film was practically identical to that of copper. Fig. 22.3 shows the morphological changes as the Ar pressure is varied between 2–10 mbar. At 2 mbar, the structure closely resembles the morphology of the thermally evaporated noble metal films. At 5 mbar, however, the large cluster aggregates, which were so characteristic of the Cu samples, are absent from the micrograph. Instead, round particles in a rather wide size range (from 3 to 30–40 nm, see Fig. 22.3 (b)) can be seen, however, most particles are not larger than 3–5 nm. At lower magnification (not shown) several aggregates of large (40–80 nm) particles can also be seen. At 10 mbar the smallest particles disappear, and the remaining ones show a log-normal type size distribution with maximum at 10 nm (Fig. 22.3 (c)).

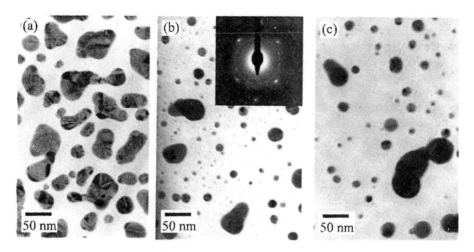

FIGURE 22.3. Morphology of the silver samples deposited onto silicon in (a): 2 mbar, (b): 5 mbar, (c): 10 mbar Ar. Reprinted from Ref.[25] with permission from Elsevier.

Analysis of the morphology of the copper and silver samples deposited at different gas pressures indicates that the location of the condensation of the particles shifted from the substrate surface to the gas phase as the pressure was elevated above a threshold value (1 mbar for Cu, 3–5 mbar for Ag).

Comparing our experimental results with the significant amount of information available in the literature about the movement of a laser ablated plume in a gas ambient,[56-68] we can give a qualitative description of the pulsed laser evaporation induced nanoparticle formation process. Accordingly, the main steps of the particle formation are as follows:

- After a few nanoseconds from the beginning of the laser pulse, ablated atoms form a very hot, dense, fastly expanding plume above the target surface.[60-62]
- If the ambient pressure is high enough, the size of the expanding, hot plume reaches the mean free path characteristic for the static ambient within a few hundreds of nanoseconds. Collisions between gas and target atoms transfer momentum to the gas, in which a supersonic shock wave develops.[60, 63] Majority of the evaporated target atoms moves confined behind the shock wave.[62-66]
- 10–100 μs after the laser pulse the evaporated target atoms become thermalized.[56-59, 63, 67] The time necessary for thermalization depends mostly on the frequency of target atom–gas atom collisions (i.e. the background gas pressure) and the efficiency of the kinetic energy exchange between them (determined by their atomic masses). At the same time – as the result of the spatial confinement – the concentration of the target atoms is still high.[62, 68]

These circumstances facilitate the nucleation of nanoclusters. After nucleation, particles grow first by absorbing the nearby target atoms and later by coalescence.

According to this model, morphology of the condensates collected on the substrates is predominantly governed by the last step of their growth, i.e. the coalescence. Coalescence can be complete, when a compact, single-crystalline particle is created or partial with necks and grain boundaries between the original particles.[69] If the process of coalescence is fast and particles encounter each other relatively rarely, complete coalescence is expected. On the other hand, high collision rate between the nanoparticles may lead to their partial coalescence. In addition, if the surface of the original particles adsorbs some contamination, e.g. oxygen, complete coalescence will be hindered and partial coalescence becomes preferred.

In case of copper, coalescence is obviously partial, while the silver particles were formed by more complete coalescence.

It should be mentioned that while the size distribution of the primarily formed nanoparticles is quite narrow, their tendency to form aggregates may hinder their application as model catalysts.

22.3.2. Nanoparticle Formation during Ion Etching of an Island Thin Film[26, 27]

As it was pointed out earlier, low energy ion bombardment can remove material from a discontinuous thin film, leading finally to formation of supported nanoparticles. A schematic representation of this idea can be seen on Fig. 22.4. Accordingly, ion etching of an island thin film consisting of particles in the 50–100 nm size range is expected to reduce all dimensions of the islands.

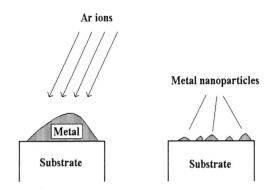

FIGURE 22.4. Scheme of nanoparticle formation during ion etching. Reprinted from Ref.[27] with permission. Copyright (2003) Springer-Verlag.

Fig. 22.5 demonstrates the realization of this idea by showing the morphological changes of a discontinuous silver layer deposited onto native oxide covered silicon during sputtering by 0.5 keV Ar^+ ions. Size distributions are shown in terms of the relative area occupied by islands of given size rather than their relative number.

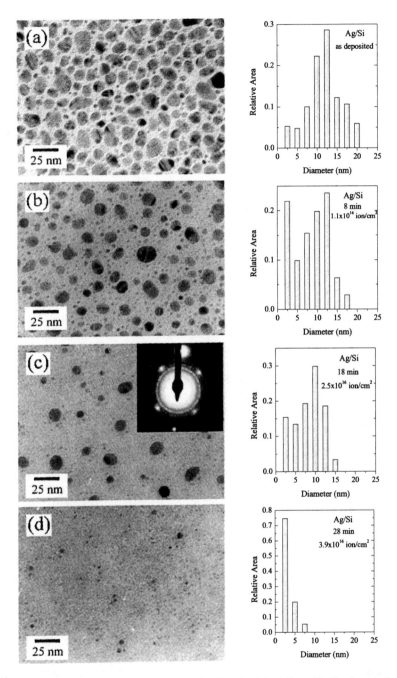

FIGURE 22.5. Morphology and particle size distribution of an island silver thin film deposited on native oxide covered silicon (a) before ion bombardment and (b) after 0.5 keV Ar$^+$ sputtering with 1.1×10^{16}, (c): 2.5×10^{16}, and (d): 3.9×10^{16} ion/cm^2 dose. Sputter velocity for silver was around 3–4 monolayers/minute. Total elapsed sputtering time is indicated on each size distribution graphs. Reprinted from Ref.[27] with permission. Copyright (2003) Springer-Verlag.

From Fig. 22.5 it can be inferred that after a short period of ion bombardment the size distribution becomes bimodal showing a large peak at the smallest sizes (Fig. 22.5 (b)). This behavior can be related to the inverse ripening phenomenon observed recently in ion implanted $SiO_2/Au/SiO_2$ structures.[48] Later the smallest particles gradually disappear (Fig. 22.5 (c)) and finally the size of the remaining islands decreases (Fig. 22.5 (d)), until the size distribution becomes unimodal, peaking at sizes below 5 nm.

According to the dark field images taken with the Ag(111) reflection, particles remain crystalline even at the end of the sputtering process. Electron diffraction indicates the presence of randomly oriented silver islands accompanied by the broad rings of ion bombardment amorphized silicon. We obtained very similar results if we used amorphous carbon substrates instead of silicon.

According to the general knowledge of ion-solid interactions, ion bombardment induced mixing is considered to be a major potential drawback of the low energy ion etching induced nanoparticle-preparation method. Although electron diffraction patterns reveal no sign of alloying, and the examples for high energy ion beam induced nanoparticle formation[43-49] suggest that it can be avoided in favorable cases, separate electron spectroscopic investigations were carried out to draw independent conclusions regarding these side effects.

The photo- and Auger-electron spectroscopic studies followed ion bombardment induced changes in different structural and bonding properties of silver films deposited onto native oxide covered Si and amorphous C substrates. The experiments, described in detail elsewhere,[27] proved unambiguously that mixing of silver and carbon, as well as silver and silicon, remains unnoticeable even at the end of sputter etching of a relatively thick film. As an example for these investigations, we present energy loss structures accompanying the most intense silver peaks for an originally 35 nm thick layer on $SiO_2/Si(100)$ and amorphous C at different stages of the ion etching process in Fig. 22.6. According to the theory of XPS, if the emitting atoms form a thin layer on the surface of the substrate, almost all of the inelastic energy loss processes must occur in the layer itself. Therefore, the energy loss structures on the low kinetic energy side of the photoelectron peaks in the layered material must resemble that observed in bulk samples. On the contrary, if emitting atoms are embedded into another phase either in the form of dissolved atoms or larger clusters, loss structures characteristic for the host phase are also expected to appear.[70]

The broad energy loss peak observed on the unbombarded $Ag/SiO_2/Si(100)$ and Ag/C samples, dissimilar in both cases to the loss structures of the substrate, is due to the hydrocarbon- and water vapor containing natural contamination layer. The easily observable loss structure due to the 1–2 nm thick overlayer confirms the sensitivity of the method for investigation of mixing effects in the nm scale. After removing the contamination layer, the characteristic loss structure accompanying the Ag 3d peaks in the $Ag/SiO_2/Si(100)$ and Ag/C samples (Fig. 22.6 (a), (b)) becomes identical to that of bulk silver. Although the intensity of the loss structures becomes very weak as silver is sputtered away, there is no indication for the appearance of the substrate plasmon loss peaks in the energy loss spectra during the course of ion etching. On the contrary, for a mixed Ag/C reference sample prepared with alternate pulsed laser evaporation of silver and graphite targets, the plasmon loss peak of carbon is very pronounced in the photoelectron spectrum of the Ag 3d core levels (Fig. 22.6 (b), spectrum (6)), in spite of the low Ag concentration (7 at%).

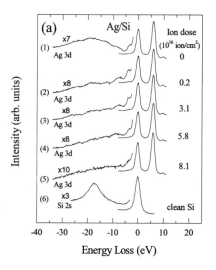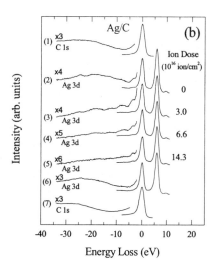

FIGURE 22.6. Core level energy loss spectra measured during ion etching of the Ag/SiO$_x$/Si (a) and the Ag/C (b) sample. The low kinetic energy region of the spectra is enlarged. Numbers on the right denote the applied ion dose. Spectrum (1) in (a) and spectra (1) and (2) in (b) correspond to the untreated samples. Spectrum (6) in (b) is due to Ag embedded in C. Spectra (6) in (a) and (7) in (b) show the characteristic loss structure of the clean substrates. Reprinted from Ref.[27] with permission. Copyright (2003) Springer-Verlag.

Although sensitivity of the applied spectroscopic methods definitely allows identification of mixing on the nm scale, our results indicate complete phase separation in the cases of both the Ag/SiO$_x$/Si and the Ag/C systems. Since it is widely acknowledged that ballistic mixing is an unavoidable consequence of ion-solid interactions, our observations can only be understood by assuming a diffusion process acting against ballistic effects.

It was shown that all kind of mixing processes accompanying ion bombardment, including the ballistic and the chemically guided atomic movement can be at least formally incorporated in a diffusion equation by proper choice of the effective diffusion coefficient.[61, 71-73] The sign of the effective diffusion coefficient can be either positive or negative. Positive coefficients lead to processes, which try to level off the concentration gradients, while negative coefficients result in uphill diffusion acting in favor of phase separation. This sign is determined by the heat of mixing of the components, emphasizing the well-documented role of the chemically biased processes[46, 71, 74-82] in the diffusion induced relaxation of ion bombarded systems. Using the formulas given by Miotello and Kelly[72, 73] or Cheng et al.[71], the effective diffusion coefficient in the Ag/C system has a high negative value. This indicates that the driving force for separation of the components in the ion bombarded sample is very pronounced even at low temperatures, in good agreement with the experimental results. The driving force for phase separation is still significant for the Ag/Si couple. Furthermore, it is also well established that diffusion coefficients in ion bombarded systems are large enough to accomplish complete phase separation over the penetration depth of keV ions.[83-87]

As a result, we can sketch a more detailed picture of the main effects accompanying ion bombardment of island thin films than we did in case of Fig. 22.4. At first, ion bombardment erodes the large islands in all three dimensions. When the penetration depth of the incoming ions reaches the thickness of the islands, atoms from the islands start to be ballistically transported into the substrate. If heat of mixing of the material of the islands and the substrate is sufficiently high and positive, chemically guided diffusional processes carry back these atoms to the islands or at least to the free surface of the sample where particle formation occurs via segregation. If the depth of ballistic mixing is increased over the range of chemically guided diffusion by elevating the ion energy, location of nanoparticle formation is expected to be shifted to deeper regions of the substrate, where chemically biased demixing processes initiate nucleation of buried particles. The structural quality of the particles is expected to be similarly high in both cases.[44-47, 49, 74, 88-91] On the contrary, if heat of mixing of the island/substrate couple has a small positive or even negative value, chemically guided diffusion will eventually enhance the ballistic mixing of the components. These results indicate that low energy ion etching of island thin films can be a valuable tool for preparation of well-controlled nanoparticles if the material of the islands and the substrate is suitably chosen.

22.4. CASE STUDIES

In the following part we extend our investigations to several physical properties of nanoparticles and their use as model catalyst, relying upon the experiences line up in the Section 22.3.

As was mentioned, thin metal films evaporated onto e.g. silicon single crystal surface can be a source of the small metal islands and may serve as model for nanoparticles. At the same time it may also serve as model catalyst because the surface of the Si(100) single crystal is always covered by native SiO_2 layer, therefore, this can be considered a $SiO_2/Si(100)$ system. The surface properties of a metal/SiO_2/Si(100) system are closely correlated with its electronic properties e.g. local density of states (LDOS).

Although most physical properties of the nanosystems – due to the extremely small amount of available material – are very difficult to study, the electronic structure of supported nanoparticles is relatively well characterized. This is mostly because of the fact that the characteristic size of the nanoparticles fits very well to the information depth of photoelectron spectroscopy, the most sensitive tool for determination of the electronic structure.

The first attempts for electron spectroscopic characterization of metal nanoparticles date back to the late seventies.[92] In the following years group 1b (Cu, Ag, Au) and group 8 metals (Ni, Pd, Pt) deposited onto a wide range of different substrates, were intensively studied.[50, 51, 93-105] However, in spite of the huge efforts, interpretation of the experimental results is still incomplete.

Generally, a gradual decrease of the valence band-width of noble metal particles with decreasing size is reported.[94-97, 100, 102, 106, 107] The effect was attributed to the decreasing interaction of d-electrons localized in neighboring atoms as the average coordination number decreases with diminishing size.[96, 108]

Broadening and shift of the metal core levels towards higher binding energies with decreasing particle sizes are a similarly general observation.[94-97, 100-102, 106, 107] The origin of the shift is, however, rather unclear. One possibility is that there is a size dependent change in the electronic structure of the particles (initial state effect[109]). Accordingly, if hybridization between the delocalized s and the localized d-states becomes stronger as the particle sizes are decreased, the core levels should shift towards higher binding energies.[94] A few optical measurements seem to support this reasoning.[110, 111] Another possibility is that the nanoparticles acquire a positive charge which is not neutralized on the time scale of the photoemission event. This charge will result in a rigid size dependent shift of all spectral elements (i.e. metal core levels, Auger lines, valence band, Fermi edge).[95, 101] Finally, it is possible that the screening of the core hole becomes less efficient in small particles as a result of increasing localization, which again leads to an apparent binding energy shift towards higher values due to a final state effect.[108] A widely used explanation for the size dependent spectral changes is combination of the last two effects.[95,96] In the smallest particles containing only 50–100 atoms (1 nm lateral size) localization can be strong enough to cause a metal-insulator transition[112, 113], which was observed by photoemission[99, 114] and scanning tunneling microscopy.[115, 116] Experiments indicating different shifts for different spectral elements,[100, 102, 103, 106, 107] however, strongly suggest that charging cannot play a significant role in the photoemission of nanoparticles.

The sample preparation method has central importance in this field. While the overwhelming majority of the studies used nanosystems utilizing island growth on the substrate, we attempted to get new insight into the electronic structural changes of nanoparticles by continuously decreasing the island sizes by low energy ion bombardment, as described above.

22.4.1. Electronic Structure of Silver Nanoparticles[53]

In noble metals the filled d-band lies 2–4 eV below the Fermi-level, while the density of states at the Fermi-energy is dominated by the free electron-like s-states. This offers the possibility of examining separately their responses to reduction of the particle size.

In Fig. 22.7 we summarize several characteristics of the core levels and the valence band of silver particles during size reduction induced by 1 keV Ar^+ ions. The particles were deposited onto Si(100) wafers covered by native SiO_2 as a nominally 2–3 nm thick discontinuous silver film. The morphological changes of the film observed by TEM were very similar to those shown in Fig. 22.5. At the end of the sputtering process the lateral size of the nanoparticles is around 5 nm, while their height is approximately 2–3 nm as estimated from the original aspect ratio of the deposited islands. All data in the figure are presented as a function of the Ag/Si atomic ratio "visible" by XPS (i. e. the ratio of the Ag $3d_{5/2}$ and Si 2s core level intensities, weighted by the corresponding atomic sensitivity factors[117]). Binding energies are referred to the Fermi edge of the grounded Cu sample holder. A rough estimation of the corresponding particle size is also given.

At the beginning of the sputtering all characteristics are identical to those measured in bulk Ag. During diminution of the particles we observed a 0.5 eV binding energy shift and a 0.3–0.4 eV broadening for the silver $3d_{5/2}$ core level. The Ag M_4VV Auger lines shifted by 0.7 eV towards lower kinetic energy. Accordingly, the Auger parameter defined as $\alpha = E_B^{Ag3d} + E_{kin}^{AgMVV}$ decreases by 0.3–0.4 eV with respect to the bulk value. The width of the 4d valence band decreases by 0.7 eV. The Ag Fermi edge measured by UPS, however, showed only a very small shift. Binding energies of the peaks belonging to the substrate did not show any appreciable shift.

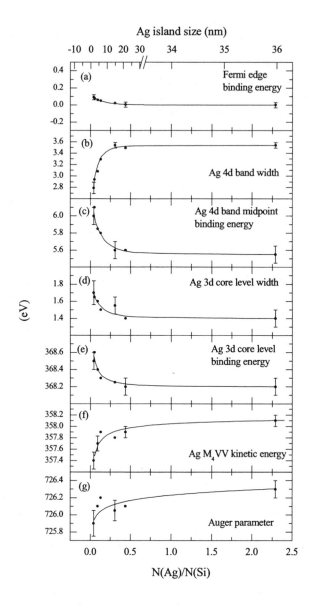

FIGURE 22.7. Spectral characteristics of Ag nanoparticles prepared on native oxide covered Si substrate. (a): binding energy of the Fermi edge, (b): full width at half maximum of the Ag 4d valence band, (c): binding energy of the midpoint of the Ag 4d band, (d): full width at half maximum of the Ag $3d_{5/2}$ core level, (e): binding energy of the Ag $3d_{5/2}$ core level, (f): kinetic energy of the Ag M_4VV Auger-transition, (g): the Auger-parameter.

We arrive at several important conclusions on the basis of the presented data. First of all, as the core level binding energy of silver and the Fermi edge shift are significantly different, any kind of charging can be excluded. This suggests that the observed changes cannot be explained by a simple final state effect. Moreover, a detailed analysis of the change of the Auger parameter as described by Jirka,[98] Kohiki[103] and Wu[118] indicates that the observed shifts can be consistently explained only by assuming changes both in the photoemission initial state and in the final state relaxation.

Thus, it is established that the most size sensitive part of the valence band in the Ag nanoparticles is the lowest binding energy peak of the 4d band. The presence of the initial state contribution suggests that there are size dependent changes in the valence band density of states of the Ag nanoparticles, which can be visualized by valence band XPS and UPS measurements.

In Fig. 22.8 we present the valence band XPS and UPS spectra of the silver nanoparticles at different stages of the size reduction process. The contribution of the substrate was subtracted. The parameter at each spectrum is the measured Ag/Si ratio.

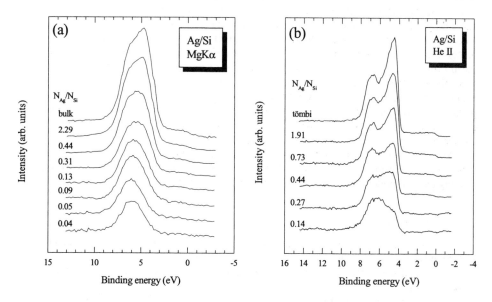

FIGURE 22.8. Valence band XPS (a) and UPS (b) spectra of silver islands on native oxide covered Si(100) during bombardment with 1 keV Ar$^+$ ions. Substrate related contributions are removed. Numbers at each spectra stand for the Ag/Si ratio determined from the appropriate XPS core level spectra. The uppermost curve is the spectrum of polycrystalline bulk Ag. Reprinted from Ref.[53] with permission from Elsevier.

The XPS valence band spectra (Fig. 22.8 (a)) are dominated by the Ag 4d band located between 3 and 8 eV. At the initial stage of sputtering ($N_{Ag}/N_{Si} = 2.29$) the shape of the 4d band is practically identical with those measured for bulk Ag. This indicates again that in the "as prepared" sample the island sizes are large enough – in agreement with the TEM data – to show bulk-like electronic structure. As sputtering proceeds, the 4d band first rounds off, losing its peak at 4.8 eV, then its maximum at 5.4 eV binding energy shifts towards higher binding energies (to 6.0 eV in the end), which is accompanied by a narrowing of the band. This is exactly what was observed for the Ag 4d band by other authors using XPS.[96, 102, 106] However, due to the limited resolution of

XPS the details in the change of the electronic structure induced by the size reduction cannot be monitored.

From the better-resolved UPS spectra (Fig. 22.8 (b)) the shift and narrowing in the 4d band observed by XPS is a consequence of the gradual reduction of the lowest binding energy peak (at 4.8 eV) in the valence band. It is interesting to note that the Ag/Si ratio at which the core level shift becomes observable is significantly lower than that at the beginning of the valence band changes, suggesting that the core level shift is a consequence of changes in the valence band density of states. This tendency seems to be in agreement with other published data.[94, 98, 100, 102, 106]

The presented results, along with those found in the literature, suggest that the observed change in the Ag 4d band cannot be interpreted in terms of a final state effect or a simple narrowing. Instead, we believe it is a collective effect of several distinct processes. For example, as the particle sizes are decreased, the contribution of the surface electronic states with different density of states compared to bulk becomes more prominent. The diminishing of the coordination number is expected to reduce the d-d and s-d hybridization and the crystal field splitting, therefore leading to narrowing of the valence d-band. On the contrary, bond length contraction (i.e. a kind of "reconstruction") is well-established in small particles.[105, 119, 120] This effect should increase the overlap of the d-orbitals of the neighboring atoms, partially restoring the width of the d-band. Nevertheless, the size dependence of these effects may be strongly different. For example, the reduction of the coordination number becomes significant only if the particles are smaller than a few hundreds of atoms.[112] Since, according to the presented data, valence band changes may begin below 10 nm particle size, it is possible that the change of the most size dependent states is governed by surface related processes.

Anyway, complete interpretation of the size dependent changes of the electronic structure of noble metal nanoparticles definitely requires further experimental work on suitably chosen model systems as well as theoretical calculations.

22.4.2. Co/Si(100) System[50, 51]

As a next step, we tried to establish a connection between the electronic structure and chemisorptive properties of supported nanoparticles. Co, a transition metal with unfilled d-band deposited onto a native SiO_2 covered Si(100) substrate was chosen as model system.

In Fig. 22.9 we show the valence band UPS spectrum of the $Co/SiO_2/Si(100)$ system.[50, 51] The spectrum of the initially deposited film is characteristic of bulk cobalt with significant Fermi level emission (top curve). When the Co film is sputtered by Ar^+ ion bombardment, the 3d valence band of Co shifts by 0.5 eV towards higher binding energies (middle curve).[50] Detailed studies unambiguously pointed out that this shift is not the result of either charging effect or cobalt silicide formation which is one of the most undesirable surface processes, but due to the reduction of particle size. This is further proven by the shift of the Co 2p core level band by 0.5 eV towards higher binding energies measured by XPS and also evidenced by the Auger parameter, which varies with the XPS core level binding energies.

In Fig. 22.10 the effect of CO chemisorption on the Co/Si(100) nanoparticles at ambient temperature is presented. The higher the dose the lower is the Co d-band intensity at the Fermi level, and the higher is the peak at around 6-7 eV which are

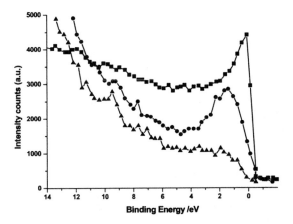

FIGURE 22.9. Valence band spectra of Co/Si(100). Upper curve: thick Co film, middle curve: Co nanoparticles, lower curve: pure Si(100)[50]

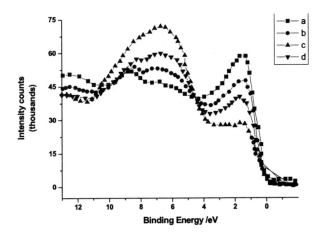

FIGURE 22.10. CO chemisorption at various dose. (a): Co without CO, (b), (c) and (d): CO chemisorbed in 1, 10 and 1000 L, respectively. Reprinted from Ref.[51] with permission from Elsevier.

characteristic of the dissociative CO chemisorption (carbon and oxygen). On the other hand, over the initially deposited Co/Si(100) layer only molecular CO chemisorption occurs at ambient temperature indicated by the two bands at 8 and 11 eV. This observation is clearly indicative of the high reactivity of nanoparticles.[51] The low temperature dissociation of the CO molecules undoubtedly illustrates the high reactivity of Co nanoparticles.

On Pt/Si(100) similar changes were observed with regards to the UPS spectra. Catalytic measurements indicated that nanoparticles of platinum have higher activity in terms of estimated turnover frequency than Pt foil.[23, 121] This is in agreement with earlier studies.[12]

22.4.3. Au/SiO₂/Si(100) system

Research on gold has significantly increased in the last decade and even today it is considered as one of the metals for future catalysis. Haruta et al. discovered exceptionally high activity of gold nanoparticles supported on Co_3O_4, Fe_2O_3 and TiO_2 oxides in CO and H_2 oxidation,[10, 122] NO reduction,[123] water-gas shift reaction,[124] CO_2 hydrogenation[125] and catalytic combustion of methanol,[126] when dispersion of the gold particles approached 100%.

Goodman and coworkers fabricated model Au/TiO₂(110) sample by epitaxially growing Ti film on Mo(110) surface. They have demonstrated how the CO oxidation rate over model Au/TiO₂(110) catalysts correlates with the thickness of gold particles and their band gap as probed by STM (this is defined as the length of the plateau in I-V plot in STS).[127, 128] The maximum activity occurs on particles that are assembled in two atomic layers thickness and at band gap of 0.2–0.6 eV. This corresponds to gold particles of ~ 5 nm in diameter.

More information is obtained from the Au $4f_{7/2}$ core level binding energy shift as a function of particle size deposited on TiO₂ and SiO₂ supports.[129] In this case the shift can be explained mainly by the final state effect. In the bulk the positive hole after photoelectron ejection is fully screened by the delocalized electrons in the valence band. On the other hand, in small nanoparticles the screening is less than in bulk, thus the kinetic energy of the measured photoelectrons is lower than that observed in the bulk, due to the Coulomb interaction between the positive hole and the photoelectrons. Consequently, the binding energy of the electrons is higher (hv = E_{kinet} + $E_{binding}$). The B.E. shift is about 0.8 eV and 1.6 eV for Au/TiO₂ and Au/SiO₂, respectively. Since the Au coverage to reach the B.E. characteristic of that of bulk value is higher on SiO₂ than on TiO₂, the difference can be interpreted by the result of the relative abilities of the metal oxide support to shield the final-state hole via extra-atomic relaxation. Thus the SiO₂ would signify a greater screening ability, i.e. TiO₂ has a larger interaction with gold.

The interaction between gold clusters and TiO₂(110) surface is also investigated via cluster nucleation and growth using STM.[130] The nucleation of gold preferentially occurs along the step edges and below 0.1 ML coverage it determines the initial deposition. Starting from the lowest coverage the Au cluster growth in 3D manner. The cluster density is independent of the surface preparation mode and of the surface defect density due to the cluster interaction with the oxygen-depleted TiO₂.

Density functional theory (DFT) calculation, however, does not fully support the strong interaction of gold with TiO₂ support.[131] In contrast to Cu and Ag, in which a strong interaction between the noble metal and the bridging oxygen was found with a transfer of the outer s-electrons of the noble metal to the Ti 3d states, Au forms weaker bonds at these sites with formation of a covalent polar bond. On Ti sites weak interaction occurs with no charge transfer. The bonding at these sites is due to metal polarization.

Campbell studied the vapor deposition of Au onto TiO₂(110) with XPS, LEED and ISS techniques.[132] The average coverage, at which the surface switches from 2D to 3D growth of Au particles, increases between 0.08±0.01 and 0.16±0.01 ML as the oxide temperature decreases from 300 to 155 K. It increased more than two-fold with increasing oxide surface defect density, induced either by mild sputtering or by annealing in vacuum. This suggests that islands nucleate at defects and that the migration of Au adatoms has unusual energetics.

Oxygen adatoms were produced on 2D and 3D gold islands on $TiO_2(110)$ using a hot filament to excite (or dissociate) O_2 gas.[133, 134] The dissociative adsorption of O_2 is thought to be impossible on pure, bulk Au surfaces. However, the unusual catalytic activity demonstrated by small Au particles on TiO_2 in e.g. CO oxidation may be related to this stronger bonded oxygen. The titration reaction of adsorbed oxygen on Au particles with CO gas ($CO_g + O_a \rightarrow CO_2$) is very rapid at room temperature, and its rate increases as island thickness increases (i.e., as the oxygen adsorption energy decreases).[133] Thin islands of Au on TiO_2 have a very weak bond to CO, so that Au sites are still free (unpoisoned by CO) at room temperature. It is possible that these bind oxygen so strongly that they can dissociatively adsorb O_2, allowing room-temperature catalytic oxidation to proceed at steady state. The thermal thickening kinetics of these Au islands have been measured using temperature-programmed ion scattering spectroscopy (TPISS).[135] Typical results show that Au island thickening begins at 300 K, but requires temperatures in excess of 900 K for completion. This very broad temperature range turns out to be very difficult to model kinetically. It suggests some unusual island energetics, whereby the activation energy for the rate-limiting step varies dramatically with the extent of sintering.

As was mentioned, reducible oxide such as TiO_2 as a support material is a key factor in developing extremely high catalytic activity. It turned out that the size of the gold nanoparticles influences their morphology (the more stable fcc crystal structure is favored in the cluster sizes n = 13–555 shown by molecular-dynamic simulation[136]). Thus, it controls the electronic structure of Au nanoparticles (e.g. band gap and B.E. shift of Au $4f_{7/2}$ band) and thereby the catalytic activity.

In order to further investigate this effect, model gold particles were deposited on a well-defined $SiO_2/Si(100)$ and a $FeO_x/SiO_2/Si(100)$ surface by pulsed laser deposition (PLD). TEM and UPS measured the morphology and electron properties, respectively. The CO oxidation was chosen as test reaction.[54]

The $Au/FeO_x/SiO_2/Si(100)$ model sample (denoted by PLD I) was depth profiled with 2 keV Ar^+ ion bombardment at a flux of $1-2 \times 10^{13}$ ion/cm^2 s. In Fig. 22.11 the valence band spectra of PLD I are compared with those recorded on $Au/SiO_2/Si(100)$. It is obvious that while on the latter sample a shift in the Au 5d valence band towards higher binding energies vs. decreasing Au/Si ratio is observed, this shift is absent when the Au/Fe ratio decreases. This means the metallic gold state is stabilized by iron oxide whereas without iron oxide a gold-silicon interaction becomes predominant.

On PLD I in as-prepared, oxidized and reduced states, the respective gold particle size was 3.8, 4.1 and 5 nm and the iron oxide support was amorphous after the first two treatments while it was partially crystallized after reduction. The activity in CO oxidation increased after oxidation of the sample, whereas it diminished after subsequent reduction. The high activity in the oxidized state was associated with amorphous iron oxide with Fe 2p B.E. = 711.3 eV. It was established that in developing catalytic activity, the gold should be metallic and the support should be amorphous with high binding energy and the reaction would occur along the perimeter of the gold particles.

In Table 22.1 the XPS results for the various states of pretreatments for both FeO_x and Au particles are presented.[54] The Au loading with regards to a few atomic layers at the outer surface of the samples is similar. After oxidation and reduction the Au 4f peaks did not show any appreciable shift being at 84.5 eV. After reductive treatment in hydrogen the decrease in the gold content is apparent. In the as-prepared state Fe 2p has

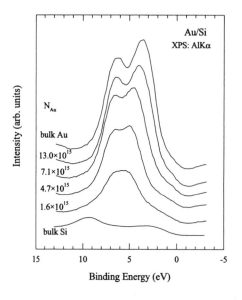
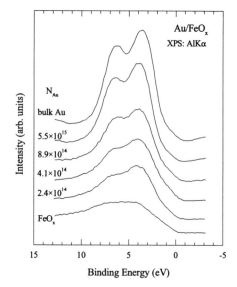

FIGURE 22.11. (a) Valence band XPS spectra of island-like gold film deposited onto silicon at different stages of the sputtering process. N_{Au} indicates the surface concentration of gold; (b) Valence band XPS spectra of island like gold film deposited onto FeO_x at different stages of the sputtering process. N_{Au} indicates the surface concentration of gold. Reprinted with permission from Ref.[54]. Copyright (2000) American Chemical Society.

TABLE 22.1. Core level characteristics of in situ pretreated PLD I sample

Treatment	Observed peaks	B.E. (eV)	Amount of Au/mg
as prepared	Au 4f	84.2	7.88×10^{-4}
	Fe 2p	710.7	
Oxidized	Au 4f	84.5	6.30×10^{-4}
	Fe 2p	711.3	
Reduced	Au 4f	84.5	5.25×10^{-4}
	Fe 2p	710.4	

a B.E. value of 710.7~710.8 eV which is assigned to Fe^{3+} state in the form of Fe_2O_3 or FeO(OH). However, significant shifts of about 1 eV can be observed for the Fe 2p peaks in the sample according to oxidation and reduction. The unchanged position of the Au peaks in the XPS spectra after oxidation and reduction pretreatments is likely due to the presence of FeO_x stabilizing the gold electronic levels, along with the decreasing surface gold concentration. The Fe 2p binding energy values change in the expected direction during pretreatments: annealing in vacuum and hydrogen somewhat reduces the iron oxides, while heat treatment in oxygen increases the level of oxidation.

TEM micrographs of the gold islands on FeO_x supported by carbon coated TEM microgrids shows the most remarkable change in the FeO_x support material.[54] In the as-prepared sample the FeO_x support gives a more or less amorphous contrast with several dark spots in a relatively broad size range. On the oxidative treatments the size of Au particles slightly increases (4.1 nm) and the size distribution of the amorphous FeO_x

becomes more homogeneous. In the electron diffraction pattern only the Au rings can be seen in both as-prepared and oxidized states. On the reduced sample, the diameter of gold further increases (5 nm) and two new FeO_x lines in the electron diffraction picture are developed. The new crystalline FeO_x phases are identified and assigned as maghemite-c, which is the cubic form of the Fe_2O_3 compared to the rhombohedral form of the Fe_2O_3 hematite. The gold size distribution becomes more symmetric.

The initial rates of CO oxidation were determined on the samples treated in different ways. The reactivity of the samples investigated in the preliminary experiments decreases in the sequence of $Au/FeO_x/SiO_x/Si(100)$ > $FeO_x/SiO_x/Si(100)$ > $Au/SiO_2/Si(100)$ > $Si(100)$. The $Au/FeO_x/SiO_2/Si(100)$ catalyst has the highest initial activity. The results along with XPS and TEM data demonstrate that the interaction of gold nanoparticles and iron oxide tends to stabilize the metallic character of gold and due to this stabilization, the Au/FeO_x catalyst has enhanced activity in the CO oxidation.

In order to explain the catalytic behavior in relation to the structure of catalyst we have to refer to other single crystal works. Single crystal studies, when metallic gold was evaporated on single crystal of TiO_2 (and not starting from Au ions),[128] indicated that the maximum in catalytic activity started when the electronic structure of gold was just at the border of the transition between ionic (large band gap) and metallic character. That is, for catalytic activity metallic gold is required, but – beyond this – a strong interaction between the metal and support along the perimeter interfaceis also necessary. The oxidized $Au/FeO_x/SiO_2/Si(100)$ sample shows the highest activity in the CO oxidation reaction. There are two reasons: (i) first, because of the oxidative removal of contaminating layer or/and (ii) the amorphous highly oxidized iron oxide is stabilized. However, the carbon content is not decisive, thus the amorphous state of iron oxide is the factor which determines the high activity of the sample.

On the other hand, the iron in the FeO_x matrix was clearly reduced, a small component in the Fe 2p spectrum due to Fe^0 appeared, while the main component was identified as Fe^{2+}. This effect may perhaps be due to the reductive effect of sputtering, but it cannot be ruled out that during catalytic treatment the whole FeO_x matrix suffered a partial reduction.

To sum up, it can be established that for developing significant catalytic activity we need (i) the presence of metallic gold particles (3–4 nm in diameter), (ii) amorphous iron oxide support in oxidized state with significant interfacing with gold particles along the perimeter, and (iii) slightly reduced iron oxide as support with oxygen defect sites.

The presence of these three factors may create the sites along the gold/support perimeter, which is a prerequisite for high catalytic activity in CO oxidation. Indeed, in the most active samples the Fe 2p lines show the same oxidation states in the XPS spectra, which could be a mixture of iron hydroxide and Fe_2O_3. TEM photographs from oxidized PLD I show an amorphous FeO_x layer. Reductive treatment clearly leads to decrease in catalytic activity, which is due to the transition from the amorphous to a partially crystalline phase and reduction of iron. The lack of catalytic activity of the PLD I sample after annealing in vacuum up to 870 K, can clearly be ascribed to the fact that a strong interaction is not present anymore. The annealing also caused reduction of the Fe particles and decreased the gold content of the surface, which led to the least reactive form of the catalyst, as also obtained by H_2 reduction.

Although there are data available in the literature which indicate the high activity of gold nanoparticles supported on non-reducible oxides, such as MgO,[137] still most data explain the activity related to gold/oxide interface. In order to separate the size and support effects and to find a correlation between the electronic properties and the catalytic activity, we try to model the pure particle size effect using inert silica support. The importance of this experiment is further underlined by the fact that within a 4–10 nm range of Au particle size the system is active,[130] but around the gold particle within a short distance the support must be defective in order to activate the reaction components. Theoretical calculations shows that 10–15 atom clusters can easily activate oxygen,[138, 139] but in real catalysts, even the smallest active ensemble consists of a few hundred atoms.

To further refine this phenomenon, a direct correlation was sought between size and electronic structure as well as activity.[140, 141] A 10 nm gold thin film was deposited by thermal evaporation in a VT-460 evaporator onto a $SiO_2/Si(100)$ wafer. The electronic structure of the Au nanoparticles was determined by UPS measuring the energy distribution of the photoelectrons excited by He (I). The samples were cleaned by Ar^+ ion bombardment for a few minutes. The gold film was ion implanted with Ar^+ ions at 4 keV and 10^{15} atom/cm^2 dose.[140] The native SiO_2 oxide layer on the Si(100) substrate served as a barrier against the Si/Au interaction, but it was thin enough to avoid electric charging. The Au 4f and Si 2p core levels were also measured to detect the binding energy and the average coverage of Au on the Si substrate. Fig. 22.12 shows the valence band UPS spectra of the sample at different stages of the ion beam treatment.

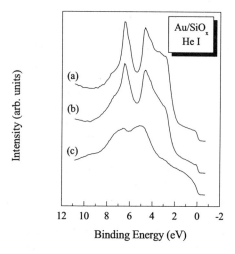

FIGURE 22.12. UPS spectra of Au/Si(100) system. After Ar ion bombardment for 0, 15, and 30 min sputtering time (spectrum (a), (b) and (c), respectively)

As illustrated in Fig. 22.12, the shape of the Au 5d valence band structure changed as indicated by a decrease in the valence bands both at 2–3 eV B.E and at 6–7 eV B.E. Finally, the whole d-valence states are redistributed below a certain size of Au nanoparticles. The separate emission from the Si/SiO_2 substrate shows that the observed effect can only be correlated with Au emission.

The UPS indicated structure change is associated with size reduction as the discontinuous gold film is transformed into rod-shape and spherical particles with sizes of 5–10 nm. Accordingly, with size reduction the activity displayed in CO oxidation was also altered: the rate increased from 6.7×10^{-3} mol min^{-1} cm^{-2} to 2×10^{-2} mol min^{-1} cm^{-2}.

Consequently, not only the gold/iron oxide interaction is responsible for the increased activity, but also the size reduction. To orchestrate this statement we assumed that if reducible FeO_x formed an interface with gold nanoparticles already tested, the catalytic activity should increase significantly due to the gold/oxide interface around the perimeter of nanoparticles.[54] Therefore, FeO_x was deposited on the Ar^+ implanted $Au/SiO_2/Si(100)$ model system that had been used several times in CO oxidation. The idea behind this investigation is that if the perimeter is active site, one should observe an effect similar to that observed earlier on $Au/FeO_x/SiO_2/Si(100)$ model system[54] regardless of the sequence of deposition of gold on iron oxide or vice versa.

In Table 22.2 the rate shows about a fourfold increase in the initial rate of CO oxidation measured on $FeO_x/Au/SiO_2/Si(100)$ as compared with $FeO_x/SiO_2/Si(100)$ and in about 60 times higher initial rate comparing to the base implanted $Au/SiO_2/Si(100)$ after catalytic reactions. These results indeed indicate that in addition to the size effect, the presence of the Au/FeO_x interface significantly influences CO oxidation. A similar effect was observed when gold was deposited on activated carbon fiber and sequentially promoted by FeO_x.[142]

TABLE 22.2. Effect of FeO_x deposited on Au particles. Reaction rate of CO oxidation

Sample	Initial rate (r_0), μmol s^{-1} cm^{-2}
$FeO_x/SiO_2/Si(100)$	2.3×10^{-2}
implanted $Au/SiO_2/Si(100)$ used in catalytic reactions	1.5×10^{-3}
implanted $Au/SiO_2/Si(100)$ after FeO_x ablation	9.5×10^{-2}

22.4.4. Pt/Carbon system[121]

Finally we wish to report a small particle fabrication method using implantation by carbon atoms with high dose (10^{17}–10^{18} atom/cm^2) and low dose (10^{15}–5×10^{16} atom/cm^2) ranges accelerated with 100 keV energy in a Van de Graaff accelerator.

Fig. 22.13 represents the UPS spectra of a sample implanted by 10^{17} atom/cm^2 carbon (Pt/C = 0.04 – 2.8). The Fig. 22.13 (a)–(f) valence bands are taken after sputtering with 0.5 keV Ar ions. In Fig. 22.13 (a) a broad valence band can be observed at higher binding energy side without any significant emission at lower binding energy that is characteristic of a carbon-covered sample. The subsequent two ion bombardments, lasting for 5 min each, result in the spectra shown in Figs. 22.13 (b) and (c). We witnessed an increased emission at low binding energy, but no intensity is at the Fermi-level. The effect of the next bombardment is shown in Fig. 22.13 (d), where Pt Fermi-edge is established and subsequent bombardments did not show any further significant variation (e)-(f). UPS results on a sample irradiated by 10^{18} atom/cm^2 (Pt/C < 0.01) shows that the Fermi-edge is not visible even after 194 min. of ion bombardments, but increased emission close to E_f is evident. At a low dose of bombardment (10^{15}–5×10^{16} atom/cm^2 range) the Pt/C ratio is high and essentially even at the shortest exposition time the Pt Fermi-edge is well developed and the shape of the valence spectra is similar to that of pure Pt. The core level of Pt 4f band along with the spin-orbit splitting and the energy gap between Pt 4f and C 1s are the same, meaning that covalent bonds are not formed between Pt and carbon.

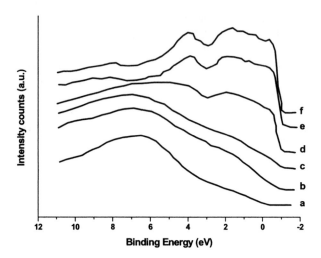

FIGURE 22.13. UPS spectra of C/Pt sample (10^{17} atom/cm^2). Ion bombardment: (a): 2 min; (b): 7 min; (c): 12 min; (d): 39 min; (e): 50 min; (f): 65 min. Reprinted from Ref.[121] with permission from Elsevier.

The He (I) excited valence band spectra are quite useful to intensify the changes. The first two Ar ion bombardments lead to increased emission close to apparent E_f, and the third ion bombardment leads to clear establishment of the Fermi-edge. The increase in height of the intensity at Fermi energy is quite rapid for the estimated data. Thus, the absence of Fermi-level emission and non-steady evolution of Pt like Fermi-edge tempts us to suggest the existence of isolated Pt particles migrating to the outermost surface of carbon. The emission before establishment of the clear Fermi-edge could be thought as of due to individual contribution of isolated Pt nanoparticles, rather than metallic valence band. This is strongly supported by Co, Cu and Au deposited on Si(100) which show a shift in valence band spectra towards higher binding energies after ion bombardment when nanoscale particles are formed.

On the sample bombarded with a higher dose of accelerated carbon atoms there is an increased Pt like emission both in X-ray and He (I) excited valence band spectra (i. e increase in emission at higher binding energy). However, no clear Fermi edge was established even after longer Ar ion bombardment. Furthermore, the Pt/C ratio on the surface did not vary significantly and was found to be close to 0.01. This might be due to a thick carbon deposit on the surface, so that even after several Ar ion bombardments, no Pt is exposed on the surface.

When the carbon flux in the Van de Graaff is lowered (< 10^{16} C atoms/cm^2), the Pt/C ratio is obviously higher. We may conclude that the UPS spectra do not indicate any Pt isolation and clear Fermi edge is established probably due to higher Pt/C surface concentration (> 0.12). The non-charge transfer interaction in Pt-C can also be estimated. It was also established that some overlapping between Pt d-orbital and carbon s-orbital occurs. The emission at Fermi level is due to this interaction between Pt and carbon and it is regarded as Pt-C non-charge transfer bonding.

22.5. GENERAL CONCLUSIONS

Nanosize metal particles play a crucial role in surface reactivity, chemisorption and catalysis. Owing to their size, the electronic structure of the particles significantly differs from that of bulk. The excess surface free energy makes the particles more reactive and structurally sensitive towards their environment.

As a consequence the modified electron band structure easily controls the bonding between the surface or/and interface of the nanoparticles and the reacting molecules (e. g. chemisorption), hence the activity and selectivity of a given catalytic reaction.

Regarding the preparation of model nanoparticles, the Chapter is focussed on the pulsed laser deposition technique and a novel approach based on material removal from existing thin film structures by low energy ion bombardment. By these techniques the change of morphology and electronic structure as a function of particle size can be monitored.

One of the major problems is the stability of the nanoparticles against coalescence, migration and restructuring during any treatments including the catalytic reaction. Short range ordering is also beneficial because active sites can be created in a manner which gives optimum performance in catalysis. Future trends concern the preparation of uniform size particles located at the surface with regular variable spacing as well as stabilization of the nanoparticles.

ACKNOWLEDGMENTS

The authors are grateful to the Hungarian Science and Research Fund for financial support (grant # T-1220, T-1372 and T-034920) and to the COST D15 program (# D15/005/99)

REFERENCES

1. G. A. Somorjai, *Chemistry of Two Dimensions Surfaces* (Cornell University Press, Ithaca, New York, 1981).
2. G. A. Somorjai, *Introduction to Surface Chemistry and Catalysis* (John Wiley & Sons, Inc. New York, 1991).
3. *Metal Cluster in Catalysis*, edited by B. C. Gates, L. Guczi and H. Knözinger, in *Stud.Surf. Sci. Catal.*, Vol. 29 (Elsevier Sci. Publ. Co., Amsterdam, 1986).
4. *Metal Nanoparticles. Synthesis, Characterization and Applications*, edited by D. L. Feldheim and C. A. Foss, Jr. (Marcel Decker, Inc., New York, 2002).
5. *Nano-Surface Chemistry*, edited by M. Rosoff (Marcel Decker Inc., New York, 2002).
6. J. Turkevich and G. Kim, Palladium: preparation and catalytic properties of particles of uniform size, *Science* **169**, 873-479 (1970).
7. L. Guczi, Precursor states in transition of carbonyl clusters to metallic particles, *Proc. 9th Int. Congress on Catalysis* edited by M. J. Phillips and M. Ternan (Chemical Institute of Canada, Ottawa, 1989), p. 114.

8. M. Che and C. O. Bennett, The influence of particle-size on the catalytic properties of supported metals, *Adv. Catal.* **36**, 55-172 (1989).

9. H. Sakurai and M. Haruta, Synergism in methanol synthesis from carbon dioxide over gold catalysts supported on metal oxides, *Catal. Today* **29**, 361-365 (1996).

10. M. Haruta, Size- and support-dependency in the catalysis of gold, *Catal Today* **36**, 153-166 (1997).

11. Y. Yuan, K. Asakura, H. Wan, K. Tsai and Y. Iwasawa, Supported gold catalysts derived from gold complexes and as-precipitated metal hydroxides, highly active for low-temperature CO oxidation, *Chem. Lett.* (9) 755-756 (1996).

12. M. Haruta, S. Tsubota, T. Kobayashi, H. Kageyama, M. J. Jenet and B. Delmon, Low-temperature oxidation of CO over gold supported on TiO_2, alpha-Fe_2O_3 and Co_3O_4, *J. Catal.* **144**, 175-192 (1993).

13. F. H. Ribeiro, A. E. Schach von Wittenau, C. H. Bartholomew and G. A. Somorjai, Reproducibility of turnover rates in heterogeneous metal catalysis: Compilation of data and guidelines for data analysis, *Catal. Rev. Sci. Eng.* **39**, 49-76 (1997).

14. T. Beutel, Z. Zhang, W. M. H. Sachtler and H. Knözinger, Temperature dependence of palladium cluster formation in NaY and 5A zeolites, *J. Phys. Chem.* **97**, 3579-3583 (1993).

15. M. Ichikawa, L. Rao, T. Ito and A. Fukuoka, Ensemble and ligand effects in selective alkane hydrogenolysis catalized on well characterized RhIr and RhFe bimetallic clusters inside NaY zeolite *Faraday Disc.* **87**, 321-336 (1989).

16. B. M. Choudary, K. Matusek, K. Lázár and L. Guczi, Fe or La promoters on the selectivity of Pd zeolites in methanol formation, *J. Chem. Soc. Chem. Comm.* (9) 592-594 (1988).

17. L. Guczi and I. Kiricsi, Zeolite supported mono and bimetallic systems: structure and performance as CO hydrogenation catalysts, *Appl. Catal. A*, **186**, 375-394 (1999).

18. I. Böszörményi, S. Dobos, K. Lázár, Z. Schay and L. Guczi, Stabilization effect of oxide supports on the decomposition of iron-ruthenium carbonyl clusters, *Surf. Sci.* **156**, 995-1002 (1985)

19. K. Lázár, K. Matusek, J. Mink, S. Dobos, L. Guczi, A. Vizi-Orosz and L. Markó, Spectroscopic and catalytic study on metal carbonyl clusters supported on Cab-O-Sil. Part I. Impregnation and decomposition of $Fe_3(CO)_{12}$, *J. Catal.* **87**, 163-178 (1984).

20. L. Guczi, A. Beck, A. Horváth and D. Horváth, From molecular clusters to metal nanoparticles, *Topics in Catalysis* **19**, 157-163 (2002).

21. M. Boutonnet, J. Kizling, R. Touroude, G. Maire and P. Stenius, Monodispersed colloidal metal particles from nonaqueous solutions - catalytic behavior in hydrogenolysis and isomerization of hydrocarbons of supported platinum particles, *Catal. Lett.* **9**, 347-354 (1991).

22. R. Touroude, P. Girard, G. Maire, J. Kizling, P. Stenius and M. Boutonnet Kizling, Preparation of colloidal platinum palladium alloy particles from nonionic microemulsions - characterization and catalytic behavior, *Colloid and Surfaces* **67**, 9-19 (1992).

23. A. S. Eppler, G. Rupprechter, L. Guczi and G. A. Somorjai, Model catalysts fabricated using electron beam lithography and pulsed laser deposition, *J. Phys. Chem., B.* **101**, 9973-9977 (1997).

24. Z. Pászti, Z.E. Horváth, G. Petõ, A. Karacs and L. Guczi, Pressure dependent formation of small Cu and Ag particles during laser ablation, *Appl. Surf. Sci.* **110**, 67-73 (1997).

25. Z. Pászti, G. Petõ, Z.E. Horváth and A. Karacs, Laser ablation induced formation of nanoparticles and nanocrystal networks, *Appl. Surf. Sci.* **168**, 114-117 (2000).

26. Z. Pászti, G. Petõ, Z. E. Horváth, O. Geszti, A. Karacs and L. Guczi, Formation of supported nanoparticles from island thin films during ion etching, *Nucl. Instr. Meth. B* **178**, 131-134 (2001).

27. Z. Pászti, G. Petõ, Z. E. Horváth, O. Geszti, A. Karacs and L. Guczi, Nanoparticle formation induced by low energy ion bombardment of island thin films, *Appl. Phys. A* **76**, 577-587 (2003).

28. P. W. Jacobs, F. H. Ribeiro, G. A. Somorjai and S. J. Wind, New model catalysts: uniform platinum cluster arrays produced by electron beam lithography, *Catal. Lett.* **37**, 131-136 (1996).

29. F. H. Ribeiro and G. A. Somorjai, The fabrication of high-technology catalysts, *Recl. Trav. Chim. Pays-Bas* **113**, 419-422 (1994).

30. P. L. J. Gunter, J. W. Niemantsverdriet, F. H. Ribeiro and G. A. Somorjai, Surface science approach to modeling supported catalysts, *Catal. Rev. Sci. Eng.* **39**, 77-168 (1997).

31. C. G. Granqvist and R. A. Buhrman, Ultrafine metal particles, *J. Appl. Phys.* **47**, 2200-2219 (1976).

32. H. Gleiter, Nanocrystalline materials, *Progr. Mat. Sci.* **33**, 223-315 (1989).

33. R. Uyeda, Studies of ultrafine particles in Japan - crystallography - methods of preparation and technological applications, *Progr. Mat. Sci.* **35**, 1-96 (1991).

34. Q. Li, T. Sasaki and N. Koshizaki, Pressure dependence of the morphology and size of cobalt(II, III) oxide nanoparticles prepared by pulsed-laser ablation, *Appl. Phys. A.* **69**, 115-118 (1999).

35. T. Koyama, S. Ohtsuka, H. Nagata and S. Tanaka, Fabrication of microcrystallites of II-VI-compound semiconductors by laser ablation method, *J. Cryst. Growth* **117**, 156-160 (1992).

36. T. Yoshida, S. Takeyama, Y. Yamada and K. Mutoh, Nanometer-sized silicon crystallites prepared by excimer laser ablation in constant pressure inert gas, *Appl. Phys. Lett.* **68**, 1772-1774 (1996).

37. A. G. Gnedovets, E. B. Kul'batskii, I. Smurov and G. Flamant, Particles synthesis in erosive laser plasma in a high pressure atmosphere, *Appl. Surf. Sci.* **96/98**, 272-279 (1996).

38. D. B. Geohegan, A. A. Puretzky, G. Duscher and S. Pennycook, Time-resolved imaging of gas phase nanoparticle synthesis by laser ablation, *Appl. Phys. Lett.* **72**, 2987-2989 (1998).

39. T. Makimura, T. Mizuta and K. Murakami, Formation dynamics of silicon nanoparticles after laser ablation studied using plasma emission caused by second-laser decomposition, *Appl. Phys. Lett.* **76**, 1401-1403 (2000).

40. M. Kaempfe, H. Hofmeister, S. Hopfe, G. Seifert and H. Graener, Morphological changes of silver nanoparticle distributions in glass induced by ultrashort laser pulses, *J. Phys. Chem. B.* **104**, 11847-11852 (2000).

41. A. L. Stepanov, D. E. Hole and P. D. Townsend, Excimer laser annealing of glasses containing implanted metal nanoparticles, *Nucl. Instr. Meth.* B **166/167**, 882-886 (2000).

42. F. Stietz, Laser manipulation of the size and shape of supported nanoparticles, *Appl. Phys. A.* **72**, 381-394 (2001).

43. F. Gonella, Nanoparticle formation in silicate glasses by ion-beam-based methods, *Nucl. Instr. Meth. B.* **166/167**, 831-839 (2000).

44. Z. Liu, H. Li, X. Feng, S. Ren, H. Wang, Z. Liu and B. Lu, Formation effects and optical absorption of Ag nanocrystals embedded in single crystal SiO_2 by implantation, *J. Appl. Phys.* **84**, 1913-1917 (1998).

45. T. Kobayashi, A. Nakanishi, K. Fukumura and G. Langouche, Fine iron particles formed in a sapphire crystal by the ion implantation technique, *J. Appl. Phys.* **83**, 4631-4641 (1998).

46. L. Thomé, J. Jagielski, G. Rizza, F. Garrido and J. C. Pivin, Formation of metallic nanophases in silica by ion-beam mixing part I: mixing mechanisms, *Appl. Phys. A.* **66**, 327-337 (1998).

47. L. Thomé, G. Rizza, F. Garrido, M. Gusso, L. Tapfer and A. Quaranta, Formation of metallic nanophases in silica by ion beam mixing part II: cluster formation, *Appl. Phys. A.* **67**, 241-247 (1998).

48. G. C. Rizza, M. Strobel, K. H. Heinig and H. Bernas, Ion irradiation of gold inclusions in SiO_2: experimental evidence for inverse Ostwald ripening, *Nucl. Instr. Meth. B.* **178**, 78-83 (2001).

49. C. J. McHargue, S. X. Ren, P. S. Sklad, L. F. Allard and J. Hunn, Preparation of manometer-size dispersions of iron in sapphire by ion implantation and annealing, *Nucl. Instr. Meth. B.* **116**, 173-177 (1996).

50. G. Pető, G. Molnár, G. Bogdányi and L. Guczi, Valence-band density-of-states of small cobalt particles on Si(111) substrate, *Catal. Lett.* **26**, 383-392 (1994).

51. G. Bogdányi, Z. Zsoldos, G. Pető and L. Guczi, CO chemisorption on small cobalt particles deposited on Si(111), *Surf. Sci.* **306**, L563-568 (1994).

52. Z. Pászti, G. Pető, Z. E. Horváth, A. Karacs and L. Guczi, Formation and valence band density of states of nanospherical Cu nanoparticles deposited on Si(100) substrate *J. Phys. Chem. B.* **101**, 2109-2115 (1997).

53. Z. Pászti, G. Pető, Z. E. Horváth, A. Karacs and L. Guczi, Electronic structure of Ag nanoparticles deposited on Si(100), *Solid State Commun.* **107**, 329-333 (1998).

54. L. Guczi, D. Horváth, Z. Pászti, L. Tóth, Z. E. Horváth, A. Karacs and G. Pető, Modeling gold nanoparticles: Morphology, electron structure and catalytic activity in CO oxidation, *J. Phys. Chem. B.* **104**, 3183-3193 (2000).

55. J. T. Cheung and H. Sankur, Growth of thin-films by laser-induced evaporation, *CRC Crit. Rev. in Solid State and Mat. Sci.* **15**, 63-109 (1988).

56. W. K. A. Kumuduni, Y. Nakayama, Y. Nakata, T. Okada and M. Maeda, Transport of YO molecules produced by Ar laser-ablation of $YBa_2Cu_3O_{7-\delta}$ in ambient oxygen gas, *J. Appl. Phys.* **74**, 7510-7516 (1993).

57. J. Gonzalo, F. Vega and C. N. Afonso, Plasma expansion dynamics in reactive and inert atmospheres during laser-ablation of $Bi_2Sr_2CaCu_2O_{(7-Y)}$, *J. Appl. Phys.* **77**, 6588-6593 (1995).

58. J. C. S. Kools, Monte-Carlo simulations of the transport of laser-ablated atoms in a diluted gas, *J. Appl. Phys.* **74**, 6401-6404 (1993).

59. T. E. Itina, W. Marine and M. Autric, Monte Carlo simulation of pulsed laser ablation from two-component target into diluted ambient gas, *J. Appl. Phys.* **82**, 3536-3542 (1997).

60. J. R. Ho, C. P. Grigoropoulos and J. A. C. Humphrey, Computational study of heat-transfer and gas-dynamics in the pulsed-laser evaporation of metals, *J. Appl. Phys.* **78**, 4696-4709 (1995).

61. R. Kelly and A. Miotello, On the mechanisms of target modification by ion beams and laser pulses, *Nucl. Instr. Meth. B* **122**, 374-400 (1997).

62. J. N. Leboeuf, K. R. Chen, J. M. Donato, D. B. Geohegan, C. L. Liu, A. A. Puretzky and R. F. Wood, Modeling of dynamical processes in laser ablation, *Appl. Surf. Sci.* **96/98**, 14-23 (1996).

63. R. Kelly, A. Miotello, B. Braren, A. Gupta and K. Casey, Primary and secondary mechanisms in laser-pulse sputtering, *Nucl. Instr. Meth. B*. **65**, 187-199 (1992).

64. Y. Tasaka, M. Tanaka, and S. Usami, Optical-emission analysis of triple-fold plume formed at pulsed ir laser-ablation of graphite, *Jpn. J. Appl. Phys.* **34**, 1673-1680 (1995).

65. R. F. Wood, J. N. Leboeuf, K. R. Chen, D. B. Geohegan and A. A. Puretzky, Dynamics of plume propagation, splitting, and nanoparticle formation during pulsed-laser ablation, *Appl. Surf. Sci.* **127/129**, 151-158 (1998).

66. F. Garrelie, C. Champeaux and A. Catherinot, Study by a Monte Carlo simulation of the influence of a background gas on the expansion dynamics of a laser-induced plasma plume, *Appl. Phys. A* **69**, 45-50 (1999).

67. H. C. Le, R. W. Dreyfus, W. Marine, M. Sentis and I. A. Movtchan, Temperature measurements during laser ablation of Si into He, Ar and O_2, *Appl. Surf. Sci.* **96/98**, 164-169 (1996).

68. A. Gupta, B. Braren, K. G. Casey, B. W. Hussey and R. Kelly, Direct imaging of the fragments produced during excimer laser ablation of $YBa_2Cu_3O_{7-\delta}$, *Appl. Phys. Lett.* **59**, 1302-1304 (1991).

69. P. B. Barna, F. M. Reicha, G. Barcza, L. Gosztola and F Koltai, Effects of co-depositing oxygen on the growth-morphology of (111) and (100) al single-crystal faces in thin-films, *Vacuum* **33**, 25-30 (1983).

70. S. Tougaard, Accuracy of the non-destructive surface nanostructure quantification technique based on analysis of the XPS or AES peak shape, *Surf. Interface Anal.* **26**, 249-269 (1998).

71. Y.-T. Cheng, Thermodynamic and fractal geometric aspects of ion-solid interactions, *Mat. Sci. Report* **5**, 45-97 (1990).

72. A. Miotello and R. Kelly, Ion-beam mixing with chemical guidance .4. Thermodynamic effects without invoking thermal spikes, *Surf. Sci.* **314**, 275-288 (1994).

73. A. Miotello and R. Kelly, Ion-beam mixing with chemical guidance .4. Thermodynamic effects without invoking thermal spikes – reply, *Surf. Sci.* **329**, 289-292 (1995).

74. C. J. McHargue, D. L. Joslin and C. W. White, Ion-beam mixing in insulator substrates, *Nucl. Instr. Meth. B* **91**, 549-557 (1994).

75. R. A. Enrique and P. Bellon, Compositional patterning in systems driven by competing dynamics of different length scale, *Phys. Rev. Lett.* **84**, 2885-2888 (2000).

76. R. A. Enrique and P. Bellon, Compositional patterning in immiscible alloys driven by irradiation, *Phys. Rev. B* **63**, 13411 (2001).

77. R. S. Averback, Atomic displacement processes in irradiated metals, *J. Nucl. Mater.* **216**, 49-62 (1994)

78. K. Nordlund, M. Ghaly and R. S. Averback, Mechanisms of ion beam mixing in metals and semiconductors, *J. Appl. Phys.* **83**, 1238-1246 (1998).

79. D. Marton, J. Fine and G. P. Chambers, Temperature-dependent radiation-enhanced diffusion in ion-bombarded solids, *Phys. Rev. Lett.* **61**, 2697-2700 (1988).

80. S. J. Simko, Y.-T. Cheng and M. C. Militello, The effects of elevated-temperature on sputter depth profiles of silver nickel bilayers, *J. Vac. Sci. Technol. A.* **9**, 1477-1481 (1991).

81. T. S. Anderson, R. H. Magruder, D. L. Kinser, R. A. Zuhr and D. K. Thomas, Formation and optical properties of metal nanoclusters formed by sequential implantation of Cd and Ag in silica, *Nucl. Instr. Meth. B* **124**, 40-46 (1997).

82. A. Meldrum, C. W. White, L. A. Boatner, I. M. Anderson, R. A. Zuhr, E. Sonder, J. D. Budai and O. Henderson, Microstructure of sulfide nanocrystals formed by ion-implantation, *Nucl. Instr. Meth. B* **148**, 957-963 (1999).

83. J. Fine, T. D. Andreadis and F. Davarya, Measurement of time-dependent sputter-induced silver segregation at the surface of a Ni-Ag ion-beam mixed solid, *Nucl. Instr. Meth.* **209/210**, 521-530 (1983).

84. D. G. Swartzfager, S. B. Ziemecki and M. J. Kelley, Differential sputtering and surface segregation - the role of enhanced diffusion, *J. Vac. Sci. Technol.* **19**, 185-191 (1981).

85. N. Q. Lam and H. A. Hoff, Surface and subsurface composition changes in Ni-Si alloys during elevated-temperature sputtering, *Surf. Sci.* **193**, 353-372 (1988).

86. D. Marton, J. Fine and G. P. Chambers, Ion-induced radiation-enhanced diffusion of silver in nickel, *Mater. Sci. Eng. A.* **115**, 223-227 (1989).

87. R. Nagel, H. Hahn and A. G. Balogh, Diffusion processes in metal/ceramic interfaces under heavy ion irradiation, *Nucl. Instr. Meth. B.* **148**, 930-935 (1999).

88. I. Sakamoto, S. Honda, H. Tanoue, N. Hayashi and H. Yamane, Structural and magnetic properties of Fe ion implanted Al_2O_3, *Nucl. Instr. Meth. B.* **148**, 1039-1043 (1999).

89. Y. S. Lee, K. Y. Lim, Y. D. Chung, C. N. Whang and Y. Jeon, X-ray absorption spectroscopy of Ag-Cr and Pd-Cr alloys formed by ion-beam-mixing, *Appl. Phys. A* **70**, 59-63 (2000).

90. D. Zanghi, A. Traverse, M. do Carmo Martins Alves, T. Girardeau and J.-P. Dallas, Structural characterization of ZrN implanted with high Co fluences, *Nucl. Instr. Meth. B* **155**, 416-425 (1999).

91. H. Hosono, Importance of implantation sequence in the formation of nanometer-size colloid particles embedded in amorphous SiO_2 - formation of composite colloids with Cu core and a Cu_2O shell by coimplantation of Cu and F, *Phys. Rev. Lett.* **74**, 110-113 (1995).

92. M. G. Mason and R. C. Baetzold, ESCA and molecular-orbital studies of small silver particles, *J. Chem. Phys.* **64**, 271-276 (1976).

93. W. F. Egelhoff Jr. and G. G. Tibbets, Growth of copper, nickel, and palladium films on graphite and amorphous-carbon, *Phys. Rev. B.* **19**, 5028-5035 (1979).

94. M. G. Mason, Electronic-structure of supported small metal-clusters, *Phys. Rev. B.* **27**, 748-762 (1983).

95. G. K. Wertheim, S. B. DiCenzo and S. E. Youngquist, Unit charge on supported gold clusters in photoemission final-state, *Phys. Rev. Lett.* **51**, 2310-2313 (1983).

96. G. K. Wertheim, S. B. DiCenzo and D. N. E. Buchanan, Noble-metal and transition-metal clusters - the d-bands of silver and palladium, *Phys. Rev. B* **33**, 5384-5390 (1986).

97. S. B. DiCenzo, S. D. Berry and E. H. Hartford Jr., Photoelectron-spectroscopy of single-size Au clusters collected on a substrate, *Phys. Rev. B* **38**, 8465-8468 (1988).

98. I. Jirka, An ESCA study of copper clusters on carbon, *Surf. Sci.* **232**, 307-315 (1990).

99. V. Vijayakrishnan, A. Chainani, D. D. Sarma and C. N. R. Rao, Metal-insulator transitions in metal-clusters - a high-energy spectroscopy study of Pd and Ag clusters, *J. Phys. Chem.* **96**, 8679-8682 (1992).

100. S. DiNardo, L. Lozzi, M. Passacantando, P. Picozzi, S. Santucci and M. DeCrescenzi, UPS and XPS studies of Cu clusters on graphite, *Surf. Sci.* **307-309**, 922-926 (1994).

101. H. Hövel, B. Grimm, M. Pollmann and B. Reihl, Cluster-substrate interaction on a femtosecond time scale revealed by a high-resolution photoemission study of the Fermi-level onset, *Phys. Rev. Lett.* **81**, 4608-4611 (1998).

102. A. R. Pennisi, E. Costanzo, G. Faraci, Y. Hwu and G. Margaritondo, Binding-energies and cluster formation at low metal-deposition - Ag on Si and SiO_2, *Physics Letters A* **169**, 87-90 (1992).

103. S. Kohiki, Photoemission from small Pd clusters on Al_2O_3 and SiO_2 substrates, *Appl. Surf. Sci.* **25**, 81-94 (1986).

104. S. V. Didziulis, K. B. Butcher, S. L. Cohen and E. I. Solomon, Chemistry of copper overlayers on zinc-oxide single-crystal surfaces - model active-sites for Cu/ZnO methanol synthesis catalysts, *J. Am. Chem. Soc.* **111**, 7110-7123 (1989).

105. M. Gautier, L. Pham Van and J. P. Duraud, Copper clusters formation on Al_2O_3 surfaces - an XPS and SEXAFS study, *Europhys. Lett.* **18**, 175-180 (1992).

106. G. Faraci, E. Costanzo, A. R. Pennisi, Y. Hwu and G. Margaritondo, Photoelectron-spectroscopy of silver clusters, *Z. Phys. D* **23**, 263-267 (1992).

107. E. Costanzo, G. Faraci, A. R. Pennisi, S. Ravesi, A Terrasi and G. Margaritondo, Photoelectron-spectroscopy of silver clusters, *Solid State Commun.* **81**, 155-158 (1992).

108. P. H. Citrin and G. K. Wertheim, Photoemission from surface-atom core levels, surface densities of states, and metal-atom clusters - a unified picture, *Phys. Rev. B.* **27**, 3176-3200 (1983).

109. J. A. Rodriguez and W. D. Goodman, Surface science studies of the electronic and chemical-properties of bimetallic systems, *J. Phys. Chem.* **95**, 4196-4206 (1991).

110. E. Anno, s-d hybridization enhancement of noble-metal particles, *Surf. Sci.* **268**, 135-141 (1992).

111. E. Anno, Optical study of continuous thin films and island films of Ir: Interband absorption and localization of conduction electrons of Ir, *J. Appl. Phys.* **85**, 887-892 (1999).

112. F. Aguilera-Granja, S. Bouarab, A. Vega, J. A. Alonso and J. M. Montejano-Carrizales, Nonmetal-metal transition in Ni clusters, *Solid State Commun.* **104**, 635-639 (1997).

113. J. Zhao, X. Chen and G. Wang, Critical size for a metal-nonmetal transition in transition metal clusters, *Phys. Rev. B* **50**, 15424-15426 (1994).

114. D. J. Huang, G. Reisfeld and M. Strongin, Photoemission study of the transition from the insulating to metallic state in ultrathin layers, *Phys. Rev. B* **55**, R1977-R1980 (1997).

115. H. N. Aiyer, V. Vijayakrishnan, G. N. Subbanna and C. N. R. Rao, Investigations of Pd clusters by the combined use of HREM, STM, high-energy spectroscopies and tunneling conductance measurements, *Surf. Sci.* **313**, 392-398 (1994).

116. C. N. R. Rao, V. Vijayakrishnan, H. N. Aiyer, G. U. Kulkarni and G. N. Subbanna, An investigation of well-characterized small gold clusters by photoelectron-spectroscopy, tunneling spectroscopy, and cognate techniques, *J. Phys. Chem.* **97**, 11157-11160 (1993).

117. *Handbook of Photoelectron Spectroscopy* edited by J. F. Moulder, W. F. Stickle, P. E. Sobol, K. D. Bomben and J. Chastain (*PHI*, Perkin-Elmer Corp. Physical Electronics Division, Eden Prairie, Minnesota, 1992).

118. Y.T. Wu, E. Garfunkel and T. E. Madey, Initial stages of Cu growth on ordered Al_2O_3 ultrathin films, *J. Vac. Sci. Technol. A* **14**, 1662-1667 (1996).

119. M. DeCrescenzi, M. Diociaiuti, L. Lozzi, P. Picoozzi and S. Santucci, Surface electron-energy-loss fine-structure investigation on the local-structure of copper clusters on graphite, *Phys. Rev. B* **35**, 5997-6003 (1987).
120. L. D. Marks, Experimental studies of small-particle structures, *Rep. Progr. Phys.* **57**, 603-649 (1994).
121. R. Sundararajan, G. Pető, E. Koltay and L. Guczi, Photoemission studies on Pt foil implanted by carbon atoms accelerated in Van de Graaff Generator: Nature of interaction between Pt and carbon, *Appl. Surf. Sci.* **90**, 165-173 (1995).
122. M. Haruta, N. Yamada, T. Kobayashi and S. J. Iijima, Gold catalysts prepared by coprecipitation for low-temperature oxidation of hydrogen and of carbon-monoxide, *J. Catal.* **115**, 301-309 (1989).
123. A. Ueda, T. Oshima and M. Haruta, Reduction of nitrogen monoxide with propene in the presence of oxygen and moisture over gold supported on metal oxides, *Appl. Catal. B.* **12**, 81-93 (1997).
124. D. Andreeva, T. Tabakova, V. Idakiev, P. Christov and R. Giovanoli, Au/alpha-Fe_2O_3 catalyst for water-gas shift reaction prepared by deposition-precipitation, *Appl. Catal. A* **169**, 9-14 (1998).
125. M. Haruta and M. Daté, Advances in the catalysis of Au nanoparticles, *Appl. Catal. A* **222**, 427-437 (2001).
126. M. Haruta, A. Ueda, S. Tsubota and R. M. Torres Sanches, Low-temperature catalytic combustion of methanol and its decomposed derivatives over supported gold catalysts, *Catal. Today* **29**, 443-447 (1996).
127. M. Valden, S. Pak, X. Lai and D.W. Goodman, Structure sensitivity of CO oxidation over model Au/TiO_2 catalysts, *Catal. Lett.* **56**, 7-10 (1998).
128. M. Valden, X. Lai and D.W. Goodman, Onset of catalytic activity of gold clusters on titania with the appearance of nonmetallic properties, *Science* **281**, 1647-1650 (1998).
129. C. C. Chusuei, X. Lai, K. Luo, and D. W. Goodman, Modeling heterogeneous catalysts: metal clusters on planar oxide supports, *Topics in Catal.* **14**, 71-83 (2001).
130. N. Spiridis, J. Haber and J. Korecki, STM studies of Au nanoclusters on $TiO_2(110)$, *Vacuum* **63**, 99-105 (2001).
131. L. Giordano, G. Pacchioni, T. Bredow and J. F. Sanz, Cu, Ag, and Au atoms adsorbed on $TiO_2(110)$: cluster and periodic calculations, *Surf. Sci.* **471**, 21-31 (2001).
132. S. C. Parker, A. W. Grant and V. T. Campbell, Island growth kinetics during the vapor deposition of gold onto $TiO_2(110)$, *Surf. Sci.* **441**, 10-20 (1999).
133. V. Bondzie, S. C. Parker and C. T. Campbell, The kinetics of CO oxidation by adsorbed oxygen on well-defined gold particles on $TiO_2(110)$, *Catal. Lett.* **63**, 143-151 (1999).
134. V. Bondzie, S. C. Parker and C. T. Campbell, Oxygen adsorption on well-defined gold particles on $TiO_2(110)$, *J. Vac. Sci. Technol. A* **17**, 1717-1720 (1999).
135. S. C. Parker, A.W. Grant, S. E. Lehto, V. Bondzie and C. T. Campbell, in preparation.
136. S. Erkoc, Stability of gold clusters: molecular-dynamics simulation, *Physica E* **8**, 210-218 (2001).
137. D. A. Cunningham, W. Vogel, H. Kageyama, S. Tsubota and M. Haruta, The relationship between the structure and activity of nanometer size gold when supported on $Mg(OH)_2$, *J. Catal.* **177**, 1-10 (1998).
138. N. Lopez and J. K. Norskov, Catalytic CO oxidation by a gold nanoparticle, *J. Am. Chem. Soc.* **124**, 11262-11263 (2002).
139. J. Guzman and B. C. Gates, Gold nanoclusters supported on MgO: Synthesis, characterizaion, and evidence of Au_6, *Nano Letters* **1**, 689-692 (2001).
140. G. Pető, G. L. Molnár, Z. Pászti, O. Geszti, A. Beck and L. Guczi, Size dependent electronic structure of gold nanoparticles deposited on SiO_x/Si(100), *Materials Sci. and Eng.* **C19**, 95-99 (2002).
141. L. Guczi, G. Pető, A. Beck, K. Frey, O. Geszti, G. Molnár and Cs. Daróczi, Gold nanoparticles deposited on SiO_2/Si(100): Correlation between size, electron structure and activity in CO oxidation, *J. Am. Chem. Soc.* **125**, 4332-4337 (2003).
142. D. A. Bulushev, L. Kiwi-Minsker, I. Yuranov, E. I. Surunova, P.A. Buffat and A. Renken, Structured Au/FeO_x/C catalysts for low-temperature CO oxidation, *J. Catal.* **210**, 149-159 (2002).

23

Nanoscale Dendrimer-Supported Hydroformylation Catalysts for Membrane Separations

M. L. Tulchinsky[*] and J. C. Hatfield[*]

23.1. INTRODUCTION

Nanoscience[1] and nanotechnology are fast growing areas whose focus is the synthesis and characterization of nanometer-scale (1 billionth of a meter) species. Application of nanotechnology in catalysis[2] implies the design of new, more efficient catalysts exhibiting nanosized features. While recent research efforts have been predominantly focused on heterogeneous catalysts, nanotechnology also might help solve challenging problems of homogeneous catalysis such as product/catalyst separation. One way to achieve this goal is to effectively increase the dimensions of small homogeneous catalytic molecules using nanoscale dendrimer supports. Dendrimers have already found use in biologic nanotechnology in developing biologically active nanodevices as smart therapeutics for cancer.[3]

Dendrimers are a special class of highly branched polymeric molecules possessing good solubility in organic solvents, mono-dispersity, well-defined nanoscale dimensions, and near-spherical shape.[4,5] An important family of these molecules with functional groups at their periphery, coined Starburst® (Polyamidoamine or PAMAM) Dendrimers, was invented at The Dow Chemical Company in the USA in the early 1980's and has been extensively patented.[6] Subsequently, scientists at DSM Fine Chemicals in the Netherlands produced another valuable class of dendrimers, DAB polypropyleneimines.[7] Both these types of dendrimers are currently available from Sigma-Aldrich in at least five different generations.

[*] The Dow Chemical Company, West Virginia Operations, 3200/3300 Kanawha Turnpike, South Charleston, WV 25303, USA.

The first reported application of dendritic ligands in homogeneous catalysis[8] appeared in 1994 along with a favorable prognosis on dendrimer catalysts functionalized at their surface.[9]   In the last five years the area has experienced rapid growth with particular emphasis on catalyst reuse and recovery.   Several recent comprehensive reviews have documented progress in this field.[10-12]

23.2. SEPARATION TECHNIQUES IN HYDROFORMYLATION

Hydroformylation offers a high level of atom economy[13] and is the largest application of homogeneous transition metal catalysis in the chemical industry.[14]   Even so, the difficulty of separating catalyst and reaction products and recycling the catalyst is especially challenging in hydroformylation where thermally sensitive organophosphorus ligands usually require delicate handling.   Several techniques have been developed for the product/catalyst separation.

23.2.1. Vaporization

In industry, the separation of low molecular weight aldehydes from hydroformylation mixtures is usually performed by vaporization or distillation.[14] However, the removal of higher molecular weight products such as C_6-C_{20} aldehydes and especially the isolation of thermally sensitive aldehydes with various functional groups by vaporization is more problematic.   Indeed, thermal operations on the corresponding mixtures can result in excessive loss of phosphorus ligands and catalysts and irreversible degradations of target aldehydes.   Overcoming these limitations is particularly important for rhodium-based catalysts and custom-made ligands because of their high cost.   Several alternative product removal techniques exist to streamline the separation of rhodium catalysts from high boiling aldehyde products.

23.2.2. Phase Separation

A two-phase aqueous hydroformylation process using water-soluble ionic phosphine ligands has been practiced commercially.[15]   This method, however, is strictly limited to feedstocks at least slightly soluble in water, such as propylene, butenes, or appropriately substituted olefins.   Furthermore, the technology requires hydrolytically stable ligands which rules out the use of highly active and regioselective organophosphites.[16]   Non-aqueous phase separation is a more general technology and allows for removal of either polar or nonpolar aldehydes after hydroformylation of higher olefins.[17-20]   However, phase separations are more difficult and may become impractical for aldehydes with intermediate polarity or when the processes such as asymmetric reactions require the use of a particular solvent.

23.2.3. "Heterogenization"

"Heterogenization" of homogeneous catalysts leads to insoluble catalysts that are recovered from the reaction mixture by simple filtration and then recycled.[21-23] Unfortunately, anchoring hydroformylation catalysts on organic or inorganic supports significantly reduces their activity due to diffusion limitations and, in addition, frequently promotes metal leaching. A heterogenized hydroformylation catalyst that suppresses metal leaching has been a rare exception.[24]

23.2.4. Membrane Separation

Membrane technology tends to replace an increasing number of traditional techniques for separating components differing in size.[25] The particle size on a macroscale or compound size on a microscale determines the type of membrane to be used. Several reviews appeared recently on reverse osmosis, nanofiltration and ultrafiltration membranes[26] along with their application in catalyst recovery and recycling.[27, 28]

For small hydroformylation ligands, the highest rhodium rejections using one-stage reverse osmosis membranes were only 95-99%.[29] Because of metal and ligand cost, rhodium catalysts require very high separation efficiencies, typically >99.9%, in a one-pass membrane unit for the process to be commercially feasible.[30] More efficient separation of rhodium complexes has been achieved using enlarged phosphines, e.g., *tris*-4-octylphenyl phosphine, and dense polymeric, nonpolar membranes.[31] The use of such modified promoters is not practical, however, due to the combined effect of their reduced reaction rate and increased molecular weight per catalytic site. Compared to triphenyl phosphine, the higher molecular weights of these ligands increase their percentage in the reaction mixture.

23.2.5. Soluble Polymeric Ligands

It has been hoped that the use of soluble supports instead of insoluble matrices would overcome sluggish activities typical for the heterogenized hydroformylation catalysts. To test this concept, high molecular weight polymeric phosphine ligands were attached to polystyrene, polyvinylchloride and polyethylene glycol. The resulting macroligands proved soluble in a variety of organic solvents, separable by membrane filtration, and competent to hydroformylate 1-pentene. Unfortunately, the recovery results were not reported.[32] Similar polymeric ligands[33] were applied in the hydroformylation of 1-decene. Although subsequent ultrafiltration afforded better than 99.8% rhodium recovery using 50 Å (5 nm) membranes,[34] the activity and regioselectivity of these ligands turned to be inferior to their low molecular weight analog.[35] In contrast, a polymeric rhodium phosphite catalyst exhibited identical catalytic activity to its parent monomeric counterpart, but this system suffered significant metal leaching in a continuous reactor.[36] A recent comprehensive review highlighted advantages and limitations of soluble polymeric ligands in many homogeneous reactions, including hydroformylation.[37]

23.3. DENDRITIC LIGANDS IN HYDROFORMYLATION

The complete functionalization of regular polymers is difficult owing to such factors as steric hindrance and limited reactivity. Furthermore, it is common for at least some fraction of the catalytically active sites to be located inside polymer chain folds and experience hindered access. Tethered catalytic sites on the periphery of a dendrimer is one way to overcome these limitations. In addition, the well-defined nanosize of a dendritic catalyst is readily adjusted for the appropriate separation technique by pre-selecting the dendrimer generation. In contrast to periphery-functionalized dendrimers, core-functionalized dendrimers[10-12, 38] have much higher molecular weight per catalytic site, potential diffusion limited mass transfer inefficiency, and problematic solubility. Neither core-functionalized nor heterogeneous dendritic catalysts[12] are discussed in this survey.

To our knowledge, the first testing of a dendritic ligand in hydroformylation was disclosed in 1996.[39] The ligand was derived from a DAB polypropyleneimine and contained 32 aminophosphite surface groups.[39] The same type of a starting dendrimer gave rise to a 32-branch multiphosphine applied in hydroformylation of 1-octene.[40] Interestingly, the hydroformylation turnover number of the latter macroligand compared well to that of the monomer.[40] Unfortunately, the efficiency of catalyst recovery was not reported in either case.

In another application, Starburst® PAMAM Dendrimer with 32 end groups (generation 3) served as the starting material for four water-soluble macroligands with hydrophilic amine or sulfonic acid groups bound to their surface.[41] The rhodium complexes incorporating these dendritic ligands catalyzed the hydroformylation of styrene and 1-octene in a two-phase toluene/water system. After the reaction, the phases were separated, but no catalyst recycling was reported. Rhodium leaching into the organic phase was in the unacceptably high range of 1.0-3.6%.

Two other groups of researchers reported diphenylphosphine-based ligands derived from synthetic carbosilane[42] or polyhedral silsesquioxane[43-45] dendrimers. In the hydroformylation of 1-octene, the former showed the same selectivity as their monomer analogs,[42] while the latter demonstrated enhanced selectivity in comparison with the monomer.[43-45] Catalyst recovery data was not released.

Although recycling of dendritic catalysts can be achieved by precipitation, two-phase separation, and immobilization on insoluble supports, it should be recognized that the same techniques are applicable to their small molecule analogs. For example, catalyst removal by phase separation[41] is readily achieved by attaching either polar[17] or nonpolar[18] groups to the hydroformylation ligand and does not require the use of a nanoscale dendritic structure. On the other hand, membrane separation relies on size differences between catalysts and products and may be the only way to achieve efficient separation when other methods fail. Nanodimensions of the catalysts significantly enhance their retention on membranes and make this approach relevant to nanotechnology. To the best of our knowledge, ours was the first reported *nanosize* catalyst recovery and recycling data by membrane filtration in homogeneous hydroformylation.[46, 47]

23.4. DESIGN AND SYNTHESIS OF NOVEL DENDROPHITES

Starburst® (PAMAM) Dendrimers contain primary amino end groups and are available in a number of generations. PAMAM generations from 0 through 4 (G0-G4) with the diameters 1.4-4.4 nm (Table 23.1) offered a convenient starting point in the construction of dendritic ligands with nanoscale dimensions. Organophosphite groups attached at the periphery of PAMAM further enlarged their size and gave rise to the name, "dendrophites." Note that phosphite end groups were selected because they promote an order-of-magnitude higher activity than phosphines.[16]

TABLE 23.1. Characteristics of PAMAM Dendrimers Containing Amino End Groups.[4]

Generation (G)	Molecular weight	Number of NH_2 groups	Size (nm)
0	517	4	1.4
1	1,430	8	1.9
2	3,256	16	2.6
3	6,909	32	3.6
4	14,215	64	4.4
5	28,826	128	5.7

The starting dendrimers contained from 4 to 64 identical primary amino end groups so that the modifying reactions[4] had to be performed on the same dendrimer molecule many times. Since surface congestion can dramatically change reaction kinetics and stoichiometry, many reactions effective on small organic molecules become significantly more difficult when applied to dendrimer exteriors.

A tailor made novel reagent coined "derivatizing phosphite" and bearing a coupling group, a spacer, and a catalytic site was designed and synthesized to introduce the phosphite catalytic sites at the dendrimer surface (Figure 23.1). A phosphine-based reagent for the derivatization of polyamines was described previously.[48-49]

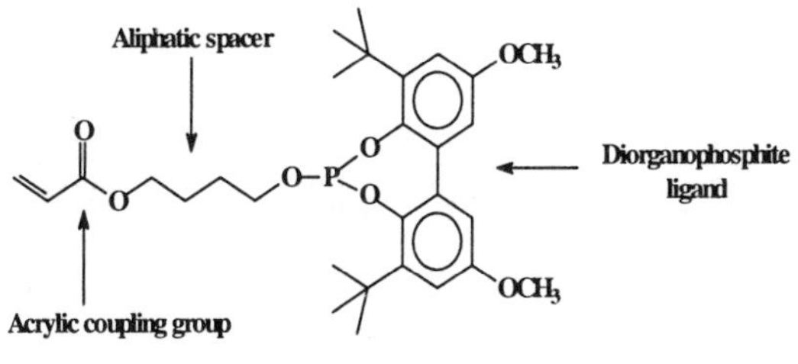

FIGURE 23.1. The "derivatizing phosphite" structure.

One essential element of the derivatizing phosphite was the acrylic moiety suitable for coupling with the surface primary amino groups of PAMAM. The Michael addition proved advantageous over other transformations to tether phosphites to dendrimers due to mild reaction conditions, quantitative yield and absence of side-products.

The second significant element, a spacer between the coupling group and the catalytic site, was necessary to prevent steric congestion on the dendrimer surface as the population of phosphite end groups increased. This phenomenon has been observed in higher generation dendrimers when surface modification was performed with bulky moieties.[50]   By increasing the distance between the phosphite groups, which occupy considerable room, and the dendrimer core, the spacer essentially amplifies the volume available for bulky terminal phosphite ligands.

Finally, the end ligand of the derivatizing phosphite can be tailor-made to suit the process and determines the catalytic properties of the species. Simple diorganophosphite groups were selected for this work due to convenient synthesis and their utility as models for more sophisticated systems. Preparation of the derivatizing reagent took only two steps including synthesis of the corresponding phosphorochloridite (Scheme 23.1).

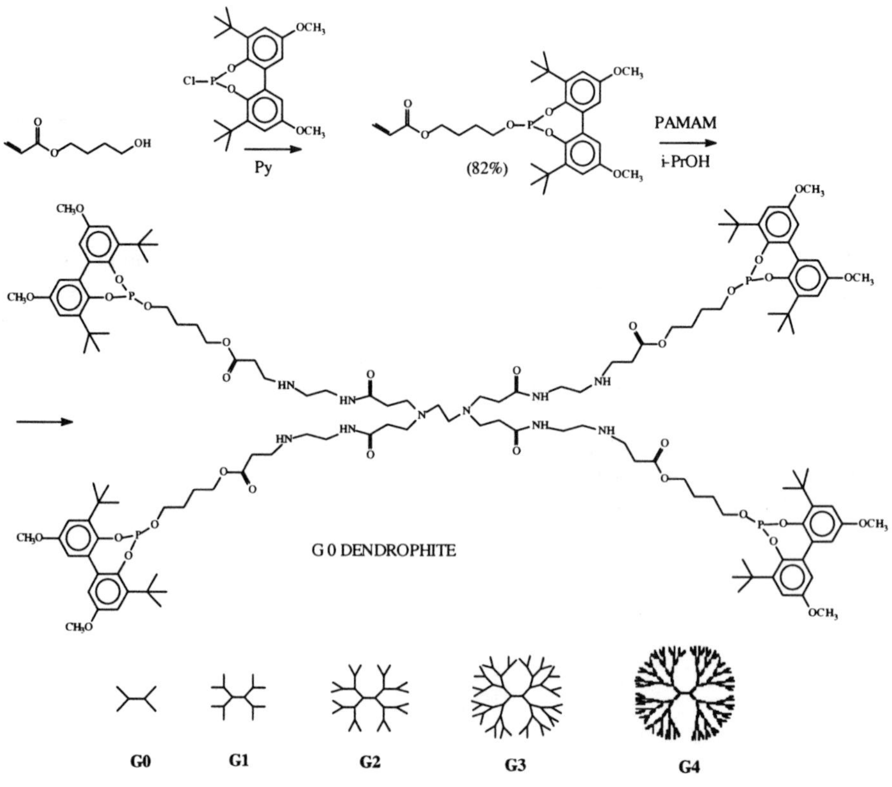

SCHEME 23.1. Dendritic multiphosphites: synthesis and graphic representation.

A Michael addition of the derivatizing phosphite to PAMAM dendrimers resulted in dendritic ligands after refluxing the two components for 16 hrs in 2-propanol. (PAMAM are soluble in water, lower alcohols and aprotic dipolar solvents such as DMF and DMSO.)   Of the two derivatizing phosphites obtained respectively from 3,3'-di-*tert*-butyl-5,5'-dimethoxy-1,1'-biphenyl-2,2'-diyl phosphorochloridite and from 3,3',5,5'-tetra-*tert*-butyl-1,1'-biphenyl-2,2'-diyl phosphorochloridite, only the former proved suitable because of the latter's poor solubility in the mentioned solvents. 2-Propanol was an acceptable solvent for both reactants (Scheme 23.1) whereas methanol reacted with the derivatizing phosphites.

The Michael addition was followed by Rimini's test with disodium nitroprusside to detect primary amino groups on the modified Starburst® (PAMAM) Dendrimers.[51] Refluxing was stopped when the test showed complete disappearance of residual primary amines. Products were isolated by concentrating the mixture with subsequent addition of hexane to precipitate white solids in 72-95% yields.   Although the method was implemented on gram scale, it is amenable to scale-up.

The dendritic ligands were characterized by elemental analysis, ^{1}H, ^{31}P, and ^{13}C NMR and FT-IR. Elemental composition of the materials attested to the addition of only one derivatizing phosphite to each primary amino group of PAMAM.   Thus, acrylic groups of the derivatizing diorganophosphites reacted with the primary amino groups of dendrimers to form 1:1 adducts. The resulting secondary amino groups resisted further reactions with derivatizing diorganophosphites presumably for steric reasons (Scheme 23.2).

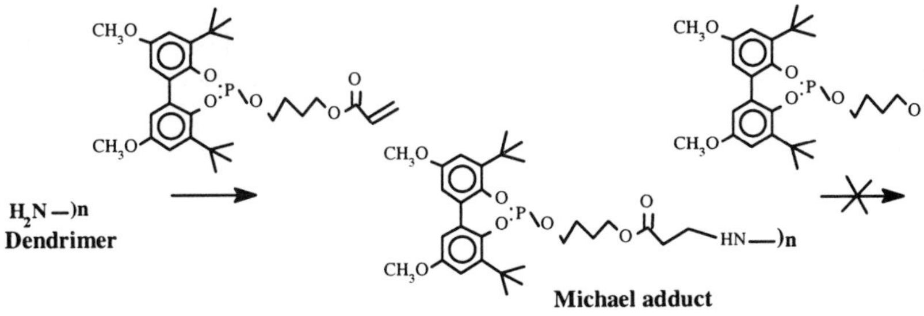

SCHEME 23.2. Addition of the derivatizing phosphite to PAMAM.

Observed ^{31}P NMR chemical shifts in the range 137.6-137.8 ppm (dmso-d$_6$) for the dendrophites are consistent with expected values (140 ppm for the monomer). Proton resonances exhibited less informative broadened lines. In contrast, all ^{13}C NMR spectra displayed sharp signals in the range 24-65 ppm which were assigned to the dendrimer core along with derivatizing phosphite aliphatic carbon atoms, 112-155 ppm – assigned to aromatic signals of diorganophosphite group, and 171-172 ppm – assigned to amide and ester carbon atoms. FT-IR spectra exhibited identifiable bands around 3,300 cm^{-1} (amide and secondary amine), 2960 (CH), 1730 cm^{-1} (ester carbonyl), 1650-1660 cm^{-1} (amide). In summary, the spectral evidence is consistent with the structure of dendritic ligands in which all primary amino groups are substituted for diorganophosphites at the macromolecule's periphery. As an example, the structure of a catalyst with 32 end phosphites is presented in Figure 23.2.

23.5. ACTIVITY, SELECTIVITY, AND STABILITY OF RHODIUM CATALYSTS WITH DENDROPHITES

Hydroformylation of propylene served as a model reaction to evaluate the novel dendrophite catalysts (Scheme 23.3). The reaction rate, which determines the process productivity, was expressed in [mol/L-h] and the reaction selectivity was measured by the ratio of normal to branched butyraldehyde isomers (N/I).

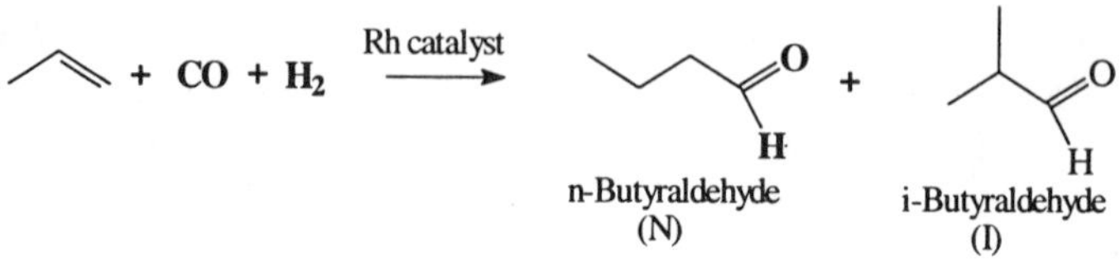

SCHEME 23.3. Model propylene hydroformylation reaction for testing the dendrophite catalysts.

The rhodium catalyst was assembled *in situ* upon mixing of rhodium (I) dicarbonyl acetylacetonate and a dendritic ligand in toluene or Texanol® under syngas pressure at 70°C. Combined dimensions of the starting PAMAM (Table 23.1) and the derivatizing phosphite (Figure 23.1) gave rise to *nanoscale* rhodium catalysts (Table 23.2, Figure 23.2).

TABLE 23.2. Activity and regioselectivity of rhodium catalysts with G0-G4 dendrophites in propylene hydroformylation.[a]

Ligand	MW[b]	Number of phosphite groups	MW per one phosphite	Estimated size (nm)	Rate, g-mol/l/h	n/i ratio
Monomeric ligand	535	1	535	<1	9.9	1.24
G0 Dendrophite	2,639	4	660	3.0	3.2	1.32
G1 Dendrophite	5,675	8	709	3.5	2.9	1.38
G2 Dendrophite	11,746	16	734	4.2	2.0	1.41
G3 Dendrophite	23,888	32	747	5.2	1.9	1.38
G4 Dendrophite	48,173	64	753	6.0	1.8	1.40

[a]Conditions: Rh 200 ppm; P/Rh = 8; 70°C; 100 psi, $C_3H_6:CO:H_2 = 1:1:1$, Texanol® solvent.
[b]Calculated based on the established structure.

It is important to note that the weight percentage of dendrophite in hydroformylation solution to maintain a certain concentration of phosphite groups has only low dependence on dendrimer generation (Table 23.2). Indeed, although the number of phosphite groups in dendrophites grows exponentially with dendrimer generation, the "effective molecular weight" or molecular weight per one phosphite group (Table 23.2) changes slowly, being only 20-40% higher than that of the corresponding monomeric diorganophosphite. Thus, the amount of dendritic promoters needed to maintain the same P/Rh ratio at fixed rhodium content is comparable to that of monomeric ligands.

A monomeric diorganophosphite obtained from the derivatizing diorganophosphite was used for comparison. As expected, an acrylic group present in this ligand inhibited hydroformylation and, thus, no activity was observed at 70°C and 100 psi syngas-propylene pressure. Therefore, the mixture was initially activated at 100°C by hydrogenation of the double bond (Scheme 23.4).

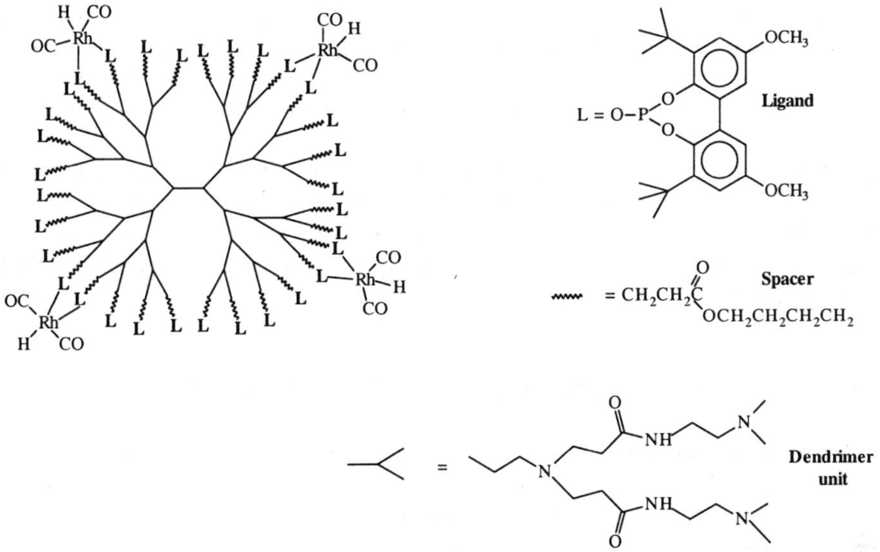

FIGURE 23.2. Structure of the Rh complex with G3 dendrophite possessing 32 terminal phosphite groups (P/Rh = 8).

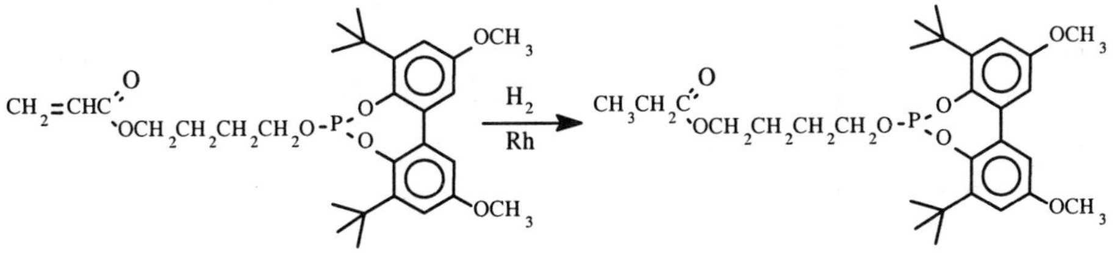

SCHEME 23.4. Activation of the monomeric diorganophosphite before hydroformylation.

The resulting saturated ligand served as a reference and the data are presented in Table 23.2. Low isomer ratios for all dendrophites are typical for diorganophosphites. The observed reaction rate for G0 was about one third that of the reference; the rate slowly declined for each subsequent generation. These effects exhibited a manifestation of the "negative dendritic effect" or decrease in catalytic activity with increasing generation number.[52] This drop, however, did not exceed 40% within the G0-G4 series. Even though catalyst activity for the phosphite-coated dendrimers was 3-5 times less than that of the unmodified diorganophosphite, it was high enough for potential commercial applications.

Hydroformylation testing with Rh/G4 dendrophite also included temperature variation at 70, 60, and 50°C. The observed temperature dependence revealed that a reaction rate was 2-3 times lower with every 10°C drop, ruling out diffusion limitations. The effect of P/Rh ratio on rate using G0 dendrophite was also studied. The initial data were obtained for P/Rh = 8. There was practically no change for P/Rh = 4, and the reaction rate decreased about 1.3 times for P/Rh = 2. No essential differences in isomer ratio were observed in any of these runs. It is important to note, however, that only P/Rh = 8 gave stable solutions. Lower P/Rh ratios led to precipitation of white solids with time, presumably owing to aggregation of dendrimers into particles with rhodium bridges.

As an indication of their stability, the catalysts showed steady activity in a continuous reactor at 70°C for one week. Dump solutions of dendritic catalysts exhibited ^{31}P NMR signals in the range of 135-136 ppm, which were assigned to the Rh complexes. Some upfield signals corresponding to small amounts (typically <5%) of ligand oxidation products were also seen.

23.6. MEMBRANE SEPARATION AND CATALYST RECYCLING

Hydroformylation can be carried out batchwise or in a continuous mode. The conceptual flow diagram for a continuous process incorporating membrane separation is depicted in Figure 23.3. Note the possibility of combining the reactor and separator as a membrane reactor to further simplify the design.

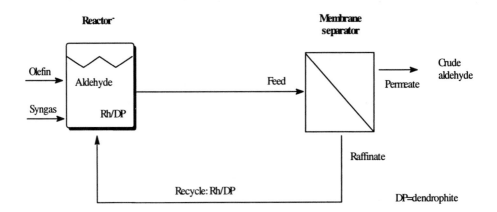

FIGURE 23.3. Continuous hydroformylation with membrane separation.

Consistent with intuition, membrane separation is enhanced as the difference in size between the transition metal complex and the product is increased.[53] While a number of nanocatalysts resulting from dendrimer modification of homogeneous catalysts have been synthesized and tested, detailed membrane studies on their separation are relatively few[27] and evaluated catalysts often suffered from low stability and metal leaching. The development of a continuous hydroformylation process combining high catalytic activity, selectivity, and stability with facile product separation and minimal catalyst leaching remains a major challenge.

Ligand testing in a batch process as reported below is a reasonable prelude to the more desirable but more complex continuous-flow membrane reactor.[28,54-56] At the same time, it is useful to recognize that batch mode production may be sufficient and even preferred for some processes, especially those resulting in high-value-added fine chemicals.

23.6.1. Reverse Osmosis Membranes

After the hydroformylations described earlier, reaction mixtures in Texanol® solvent were passed through reverse osmosis membranes MPF-50 (Membrane Products Kiryat Weizmann Ltd., Israel) or cross-linked GKSS membranes having active layer thicknesses of 1 μm and 10 μm (GKSS, Forschungszentrum Geesthach GmbH, Germany). All tests were conducted in a nitrogen environment at room temperature. All three membranes proved resistant to Texanol® and the aldehydes in preliminary tests. Also, no detectable dendritic material was found in the permeate solutions, indicating molecularly robust dendrophites under these conditions. To quantify membrane performance, rhodium rejection was determined according to the formula:

Rejection = [1 - Permeate Conc./0.5(Feed Conc. + Raffinate Conc.)]x 100%

The total volume of permeate collected from each membrane over a specified period was used to calculate the permeate flux in gallons/foot^2/day (GFD).

TABLE 23.3. Rhodium rejection and flux rate for reverse osmosis membranes challenged by hydroformylation mixtures[a] containing dendrophites.

Ligand (dendrimer generation)	MPF-50		GKSS (1μm)		GKSS (10 μm)	
	% Rh rejection	Flux rate (GFD[b])	% Rh rejection	Flux rate (GFD[b])	% Rh rejection	Flux rate (GFD[b])
G0	99.59	0.15	99.60	0.70	99.84	0.20
G1	99.74	0.34	99.86	0.52	99.86	0.30
G2	99.88	0.11	99.95	0.14	99.96	0.18
G3	99.92	0.10	99.94	0.24	99.96	0.24
G4	99.89	0.063	99.93	0.063	99.94	0.067
Ligand A	65	0.3-0.5	-	-	88	0.15-0.2

[a] Conditions: p = 300 psig for MPF-50 and GKSS (1μm); p = 150 psig for GKSS (10 μm); Texanol® solvent.
[b] GFD = gallons/foot^2/day.

Table 23.3 summarizes Rh rejections and permeation rates for reverse osmosis membranes challenged by five dendrophite generations. The same test was also performed using the low molecular weight reference ligand, Ligand A (Figure 23.4), which, consistent with expectations owing to its much lower molecular weight, showed much lower rejection compared to the dendrophites.

23.6.2. Ultrafiltration (UF) Membranes

As seen in the previous section, dendrophite catalysts were separated from aldehyde products with excellent rhodium recovery up to 99.96% using reverse osmosis membranes. Although the high rhodium rejection by these membranes was close to commercial requirements, their low permeation rates disqualified them from serious consideration.

In contrast, UF membranes typically operate at lower pressures (30-80 psig) than reverse osmosis membranes (150-1,000 psig) and their fluxes often are an order of magnitude higher.[26] The separation experiments in this work were performed with G2-G4 dendrophite catalysts employing 50 Å (5 nm) ultrafiltration membranes (US Filter/Membralox, Warrendale, PA). The ultrafiltration set-up is shown in Figure 23.4.

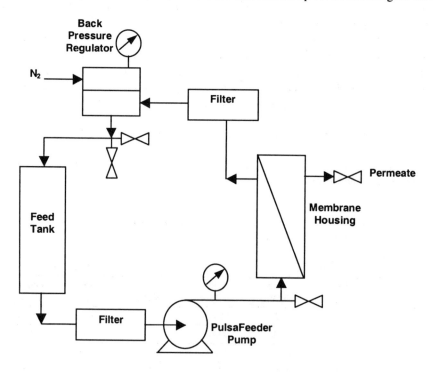

FIGURE 23.4. Apparatus for separation of hydroformylation mixtures using a T170 50 A (5 nm) ceramic membrane.

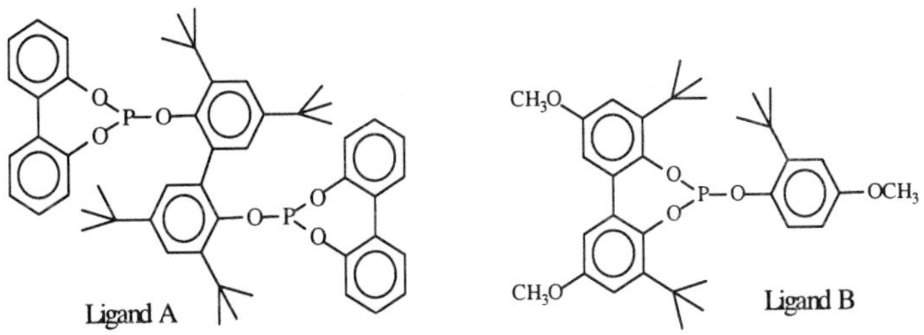

FIGURE 23.5. Structure of reference monomeric ligands. Ligand A is referred to the preceding section and Ligand B is discussed in this section.

The G2-G4 dendrophite generations provided the catalytic species having molecular diameters of around 5 nm (see Table 23.2). These catalysts intended for nanoscale separation experiments were prepared in toluene (Rh ~ 100 ppm, P/Rh = 8) and their performance was evaluated using the propylene hydroformylation model reaction. As mentioned above, the high P/Rh ratio was necessary to stabilize the catalyst in the membrane testing; use of lower ratios led to a gradual precipitation of solids. A monodentate phosphite ligand (Figure 23.5, Ligand B) was used for comparison. Table 23.4 shows that both higher rejections and permeation rates were obtained for the UF membrane compared to the reverse osmosis membranes. It is also seen that three consecutive nanoscale separations with G4 dendrophite exhibited every time very high membrane rejections of the catalyst and high flux rates.

Recycling experiments were performed using the second (G2) and third (G3) generation dendrophites. Following propylene hydroformylation in toluene, rhodium complexes with dendrophites were separated from the reaction mixture by ultrafiltration and then the catalyst solutions were used in a second hydroformylation reaction. Rate differences between the two runs were only 2% for the G2 and 3% for the G3 dendrophites. In all of the experiments, ^{31}P NMR analyses of recycled catalyst solutions confirmed integrity of the G2, G3, and G4 dendrophites.

TABLE 23.4. Permeate rate and Rhodium rejection for membrane ultrafiltration of hydroformylation mixtures.[a]

Dendrimer generation	Cycle	% Rh rejection	Flux rate (GFD)
G2	1st	99.994	2.0
G3	1st	99.997	1.9
G4	1st	99.997	1.3
G4	2nd	99.997	2.2
G4	3rd	99.994	2.1
Ligand B	1st	83-88	4.2

[a] Conditions: p = 50 psig; toluene solvent · [b] GFD = gallon/foot^2/day.

23.6.3. Suppression of Metal Leaching

Metal leaching is a significant problem for polymeric catalysts.[36] Its chemical nature presumably involves an equilibrium between macroligand-bound and "free" rhodium:

$$\text{Rh} + \text{P} \rightleftharpoons \text{Rh-P}$$

"free"   ligand-bound

In contrast to nanoscale complexes with dendrimers, "free" rhodium species have small, Angstrom-scale dimensions and apparently pass through the membrane with the product. This leached rhodium is no longer available to bind with polymeric ligands and the active catalyst concentration is reduced. In the prior art, at least 0.2% rhodium remained with the product following ultrafiltration separation of polymeric hydroformylation ligands.[34]

In our work, rhodium leaching during the separation of aldehyde product mixtures from higher dendrimer-based catalysts was negligible (see Tables 23.3, 23.4). The level of rhodium in the product (permeate) was extremely low for the UF membrane, less than 0.01%. This practical absence of metal leaching in the case of higher dendrophites was an unexpected and propitious phenomenon and may have at least two reasons:

(1) Surface phosphite groups of the nanosized dendritic ligands were located in close proximity to one another and may have chelated rhodium species so tightly as to prevent their leaching. This capacity of dendritic ligands to tightly bind metal species is apparently similar to additive ligands in cases of reduced syngas pressure and in the absence of syngas.[57,58] Note that the absence of strong rhodium coordination with phosphites in regular polymeric ligands may be due to the haphazard location of organophosphorus sites throughout the polymer structure.

(2) Functional groups such as amino, amide, and ester located throughout the dendrimer framework also can coordinate with rhodium. These functionalities can bind rhodium under the syngas-deficient conditions during separation and help prevent its leaching into the product.

23.7. CONCLUSIONS AND OUTLOOK

This work demonstrates how the application of nanotechnology principles in homogeneous catalysis uncovers new and practical solutions to the complex problem of product/catalyst separation. Use of nanoscale dendritic ligands in hydroformylation specifically and in homogeneous catalysis generally is an elegant approach to the on-going problem of catalyst recovery. The very high rhodium rejections achieved in membrane separations of hydroformylation mixtures using rhodium catalysts with novel nanosized dendrophite ligands exceeded the recovery goal. Flux rates for commercial UF membranes were an order of magnitude larger than those for reverse osmosis membranes and proved close to match commercial requirements. Separation efficiencies for the nanoscale G2-G4 dendrophites were uniformly high and suggested the practical use of less expensive, lower generation dendrimers (such as G2) for hydroformylation catalysts.

Additional experiments will be necessary to evaluate dendrophites in a continuous process. Furthermore, the relatively high cost of dendrimers will require high catalyst turnover numbers to justify the investment. Usually there is no single solution to the formidable challenge of catalyst separation. It is too early to determine if dendritic catalysts will be broadly utilized or more appropriate for niche applications.

ACKNOWLEDGEMENT

Mr. Heqi Pan is acknowledged for the synthesis and collecting data on dendrophite catalytic activity. The thorough and diligent work of Peter Ng and Tom Baldy in assembling, operating, and continuously improving the membrane apparatuses is greatly appreciated. We also acknowledge Marina Dumer for her help in matching the format requested by the publisher for this chapter. Finally, we thank The Dow Chemical Company for the permission to publish this work.

REFERENCES

1. S. A. Borman, Nanoscience, *Chem. Eng. News* **80** (50), 46 (2002).
2. M. Jacoby, Nanosized Catalysts, *Chem. Eng. News* **80** (37), 30-32 (2002).
3. See the website: http://nano.med.umich.edu/Dendrimers.html
4. D. A. Tomalia, A. M. Naylor, and W. A. Goddard III, Starburst Dendrimers: Molecular-Level Control of Size, Shape, Surface Chemistry, Topology, and Flexibility from Atoms to Macroscopic Matter, *Angew. Chem. Int. Ed. Eng.* **29** (2), 138-175 (1990).
5. G. R. Newkome, C. N. Moorefield, and F. Voegtle, Dendritic Molecules. Concepts, Syntheses, Perspectives, Weinheim: VCH, 1996, 250 pp.
6. D. A. Tomalia et al. US Patents Nos. 4,435,548; 4,507,466; 4,558,120; 4,568,737; 4,587,329; 4,599,400; 4,631,337; 4,690,985; 4,694,064; 4,713,975; 4,737,550; 4,871,779; 4,857,599; 5,041,516; 5,338,532; 5,387,617; 5,393,795; 5,393,797; 5,527,524; 5,560,929; 5,714,166; 5,731,095; 5,773,527; 5,919,442; 6,020,457; 6,043,336; 6,177,414; 6,224,898; 6,312,679; 6,471,968; 6,475,994.
7. E. W. Meijer, H. J. M. Bosman, F. H. A. M. J. Vandenbooren, E. M. M. De Brabander-van den Berg, A. M. C. F. Castelijns, H. C. J. De Man, R. W. E. G. Reintjens, C. J. C. Stoelwinder, and A. J. Nijenhuis, Dendritic Macromolecules and the Preparation Thereof, US Patent No. 5,530,092, June 25, 1996.
8. J. W. J. Knapen, A. W. van der Made, J. C. de Wilde, P. W. N. M. van Leeuwen, P. Wijkens, D. M. Grove, and G. van Koten, Homogeneous Catalysts Based on Silane Dendrimers Functionalized with Arylnickel(II) Complexes, *Nature* **372**, 659-663 (1994).
9. D. A. Tomalia and P. R. Dvornic, What Promise for Dendrimers? *Nature* **372**, 617-618 (1994).
10. G. E. Oosterom, J. N. H. Reek, P. C. J. Kamer, and P. W. N. M. van Leeuwen, Transition Metal Catalysis Using Functionalized Dendrimers, *Angew. Chem. Int. Ed.* **40** (10), 1828-1849 (2001).
11. D. Astruc and F. Chardac, Dendritic Catalysts and Dendrimers in Catalysis, *Chem. Rev.* **101** (9), 2991-3024 (2001).
12. R. van Heerbeek, P. C. J. Kamer, P. W. N. M. van Leeuwen, and J. N. H. Reek, Dendrimers as Support for Recoverable Catalysts and Reagents, *Chem. Rev.* **102** (10), 3717-3756 (2002).
13. B. M. Trost, The Atom Economy – A Search for Synthetic Efficiency, *Science* **254** (5037), 1471-1474 (1991).
14. C. D. Frohning and C. W. Kohlpaintner in Applied Homogeneous Catalysis with Organometallic Compounds, Eds. B. Cornils and W. A. Herrmann, vol.1, VCH: Weinheim, 1996, pp. 27-104.
15. B. Cornils and E. Kunz in Aqueous Phase Organometallic Catalysis, Concepts and Applications, Eds. B. Cornils and W. A. Herrmann, Willey-VCH, New York, 1998, pp.273-279.
16. E. Billig, A. G. Abatjoglou, and D. R. Bryant, Bis-Phosphite Compounds, US Patent No. 4,885,401, December 5, 1989.
17. J. N. Argyropoulos, D. R. Bryant, D. L. Morrison, K. E. Stockman, and A. G. Abatjoglou, Separation Processes, US Pat. No. 5,932,772, August 3, 1999.
18. J. N. Argyropoulos, D. R. Bryant, D. L. Morrison, and K. E. Stockman, Separation Processes, US Pat. No. 5,952,530, September 14, 1999.
19. E. E. Bunel, Process of Preparation of Linear Aldehydes, US Pat. No. 6,175,043, January 16, 2001.

20. P. Arnoldy, C. M. Bolinger, E. Drent, J. van Gogh, C. H. M. van der Hulst, and R. Moene, Hydroformylation Process, US Pat. No. 6,187,962, February 13, 2001.

21. P. Panster and S. Wieland in ref. 14, pp. 605-623.

22. B. P. Santora and M. R. Gagne, A Wolf in Sheep's Clothing, *Chem. Innovation* **30** (8), 22-29 (2000).

23. C. A. McNamara, M. J. Dixon, and M. Bradley, Recoverable Catalysts and Reagents Using Recyclable Polystyrene-Based Supports, *Chem. Rev.* **102** (10), 3275-3300 (2002).

24. A. J. Sandee, L. A. van der Veen, J. N. H. Reek, P. C. J. Kamer. M. Lutz, A. L. Spek, and P. W. N. M. van Leeuwen, A Robust, Environmentally Benign Catalyst for Highly Selective Hydroformylation, *Angew. Chem. Int. Ed.* **38** (21), 3231-3235 (1999).

25. Industrial Membrane Separation Technology, Eds. K. Scott, R. Hughes, London: Blackie Academic & Professional, **1996**, 305 pp.

26. R. Singh, Industrial Membrane Separation Processes, *CHEMTECH* **28** (4), 33-44 (1998).

27. H. P. Dijkstra, G. P. M. van Klink, and G. van Koten, The Use of Ultra- and Nanofiltration Techniques in Homogeneous Catalyst Recycling, *Acc. Chem. Res.* **35** (9), 798-810 (2002).

28. I. F. J. Vankelecom, Polymeric Membranes in Catalytic Reactors, *Chem. Rev.* **102** (10), 3779-3810 (2002).

29. J. F. Miller, D. R. Bryant, K. L. Hoy, N. E. Kinkade, and R. H. Zanapalidou, Membrane Separation Process, US Patent No. 5,681,473, October 28, 1997.

30. N. Brinkmann, D. Giebel, G. Lohmer, M. T. Reetz, and U. Kragl, Allylic Substitution with Dendritic Palladium Catalysts in a Continuously Operating membrane Reactor, *J. Catal.* **183**, 163-168 (1999).

31. H. W. Deckman, E. Kantner, J. R. Livingston, Jr., M. G. Matturro, and E. J. Mozelski, Method for Separating Catalyst from a Hydroformylation Reaction Product using Alkylated Ligands, US Patent No. 5,395,979, March 7, 1995.

32. E. Bayer and V. Schurig, Soluble Metal Complexes of Polymers for Catalysis, *Angew. Chem. Int. Ed. Eng.* **14** (7), 493-494, (1975).

33. N. S. Imyanitov, V. A. Rybakov, S. B. Tupitsyn, and V. I. Gorchakova, Synthesis of High-Molecular Weight Phosphine Ligands for Catalysts of Homogeneous Reactions, *Neftekhimia* **32** (3), 200-207, (1992).

34. S. B. Tupitsyn and N. S. Imyanitov, Separation of Rhodium Catalysts with Polymeric Phosphine Ligands for Homogeneous Hydroformylation by the Ultrafiltration Method, *Neftekhimia* **36** (3), 249-254, (1996).

35. N. S. Imyanitov, V. A. Rybakov, S. B. Tupitsyn, S. V. Zubarev, and N. A. Alekseeva, Development of Novel Hydroformylation Technology, Khim. Prom-st. (Moscow), (11), 649-651 (1991); CA116: 154177p.

36. T. Jogsma, P. Kimkes, G. Challa, and P. W. N. M. van Leeuwen, A New Type of Highly Active Polymer-bound Rhodium Hydroformylation Catalyst, *Polymer* **33** (1), 161-165 (1992).

37. D. E. Bergbreiter, Using Soluble Polymers to Recover Catalysts and Ligands, *Chem. Rev.* **102** (10), 3345-3384 (2002).

38. G. E. Oosterom, S. Steffens, J. N. H. Reek, P. C. J. Kamer, and P. W. N. M. Van Leeuwen, Core-functionalized Dendrimeric Mono- and Diphosphine Rhodium Complexes: Application in Hydroformylation and Hydrogenation, *Top. Catal.* **19** (1), 61-73 (2002).

39. E. Wissing, A. J. J. M. Teunissen, C. B. Hansen, P. W. N. M. Van Leeuwen, A. Van Rooy, and D. Burgers, Process for the Preparation of an Aldehyde, WO 96/16923, June 6, 1996.

40. M. T. Reetz, G. Lohmer, and R. Schwickardi, Synthesis and Catalytic Activity of Dendritic Diphosphane Metal Complexes, *Angew. Chem. Int. Ed. Engl.* **36** (13/14), 1526-1529 (1997).

41. A. Gong, Q. Fan, Y. Chen, H. Liu, C. Chen, and F. Xi, Two-phase Hydroformylation Reaction Catalyzed by Rhodium-complexed Water-soluble Dendrimers, *J. Mol. Catal. A* **159**, 225-232 (2000).

42. D. De Groot, P. G. Emmerink, C. Coucke, J. N. Reek, P. C. J. Kamer, and P. W. N. M. van Leeuwen, Rhodium catalysed hydroformylation using diphenylphosphine functionalised carbosilane dendrimers, *Inorg. Chem. Commun.* **3** (12), 711-713 (2000).

43. D. De Groot, P. G. Emmerink, C. Coucke, J. N. Reek, P. C. J. Kamer, and P. W. N. M. van Leeuwen, Rhodium catalysed hydroformylation using diphenylphosphine functionalised carbosilane dendrimers, *Inorg. Chem. Commun.* **3** (12), 711-713 (2000).

44. L. Ropartz, R. E. Morris, D. F. Foster, and D. J. Cole-Hamilton, Phosphine-containing Carbosilane Dendrimers Based on Polyhedral Silsesquioxane Cores as Ligands for Hydroformylation Reaction of Oct-1-ene, *J. Mol. Catal. A* **182-183**, 99-105 (2002).

45. L. Ropartz, R. E. Morris, D. F. Foster, and D. J. Cole-Hamilton, Increased selectivity in hydroformylation reactions using dendrimer based catalysts; a positive dendrimer effect, *Chem. Commun.* (4), 361-362 (2001).

46. L. Ropartz, R. E. Morris, G. P. Schwartz, D. F. Foster, and D. J. Cole-Hamilton, Dendrimer-bound tertiary phosphines for alkene hydroformylation,*J. Inorg. Chem. Commun.* **3** (12), 714-717 (2000).

47. M. L. Tulchinsky and D. J. Miller, Dendritic Macromolecules for Metal-Ligand Catalyzed Processes, US patent No. 6,350,819, February 26, 2002.
48. M. L. Tulchinsky and D. J. Miller, Dendritic Macromolecules for Metal-Ligand Catalyzed Processes, US patent No. 6,525,143, February 25, 2003.
49. P. Lange, A. Schier, and H. Schmidbaur, Mono-, Di- and Trinuclear Gold(I) Complexes of New Phosphino-Substituted Amides: Initial Steps toChlorogold(I)diphenyl-phosphino-terminated Dendrimers, *Inorg. Chim. Acta* **235** (1-2), 263-272 (1995).
50. P. Lange, A. Schier, and H. Schmidbaur, Dendrimer-Based Multinuclear Gold(I) Complexes, *Inorg. Chem.* **35** (3), 637-642 (1996).
51. J. F. G. A. Jansen, E. M. M. De Brabander-van den Berg, and E. W. Meijer, Encapsulation of Guest Molecules into a Dentritic Box, *Science* **266** (5188), 1226-1229 (1994).
52. B. S. Furniss, A. J. Hannaford, P. W. G. Smith, and A. R. Tatchell, Vogel's Textbook of Practical Organic Chemistry, Longman Scientific & Technical: London, 1989, p. 1217.
53. A. W. Kleij, R. A. Gossage, J. T. B.H. Jastrzebski, J. Boersma, and G. van Koten, The Dendritic Effect in Homogeneous Catalysis with Carbosilane-Supported Arylnickel(II) Catalysts: Observation of Active-Site Proximity Effects in Atom-Transfer Radical Addition, *Angew. Chem. Int. Ed. Engl.* **39** (1), 176-178 (2000).
54. A. Behr in Aqueous Phase Organometallic Catalysis, Concepts and Applications, Eds. B. Cornils, W. A. Hermann, Willey-VCH: New York, 1998, p. 172.
55. K. K. Sirkar, P. V. Shanbhag, and A. S. Kovvali, Membrane in a Reactor: A Functional Perspective, *Ind. Eng. Chem. Res.* **38** (10), 3715-3737 (1999).
56. J. Woltinger, A. S. Bommarius, K. Drauz, and C. Wandrey, The Chemzyme Membrane Reactor in the Fine Chemicals Industry, *Org. Proc. Res. Dev.* **5** (3), 241-248 (2001).
57. H. P. Dijkstra, C. A. Kruithof, N. Ronde, R. van de Coevering, D, J. Ramon, D. Vogt, G. P. M. van Klink, and G. van Koten, Shape-Persistent Nanosize Organometallic Complexes: Synthesis and Application in a Nanofiltration Membrane Reactor, *J. Org. Chem.* **68** (3), 675-685 (2003).
58. D. R. Bryant and T. W. Leung, Hydroformylation Process Employing Indicator Ligands, US patent No. 5,741,943, April 21, 1998.
59. D. R. Bryant and T. W. Leung, Process Employing Indicator Ligands, US patent No. 5,741,945, April 21, 1998.

24

The Catalytic/Oxidative Effects of Iron Oxide Nanoparticles on Carbon Monoxide and the Pyrolytic Products of Biomass Model Compounds

P. Li, E. J. Shin, D. E. Miser, F. Rasouli, and M. R. Hajaligol*

24.1. INTRODUCTION

The formation of CO has been of concern in many industrial processes such as incineration, automotive exhausts, coal and biomass combustion, and the burning of tobacco. Transition metal oxides are desirable alternatives to the precious metals based CO catalysts for many of these applications since they are less costly. Nanophase transition metal oxides, typically with a particle size of less than 100 nm, could potentially provide significantly improved catalytic performance over micron size materials, due to their high surface area and more densely populated surface unsaturated sites. The advancement of nanotechnology has made many nanophase transition metal oxides available. The material studied here is a substance termed "NANOCAT® Superfine Iron Oxide (Fe_2O_3)" by its manufacturer, MACH I Inc., of King of Prussia, PA. The material (referred to simply as NANOCAT® hereafter), with an average particle size of 3 nm, was first studied about ten years ago by Huffman et al[1] as a potential catalyst for the direct coal liquefaction (DCL) process. The subsequent papers by Zhao et al[2] and Zhen et al[3] characterized its structure and reported a phase transition on annealing.

* P. Li, E. J. Shin, D. E. Miser, F. Rasouli, and M. R. Hajaligol, Philip Morris USA, Research Center, Richmond, Virginia, 23234.

Although iron oxide has been known as a CO catalyst for some time,[4] NANOCAT® had not been evaluated as a CO catalyst until very recently (Li *et al*[5]). It was found that, compared to other iron oxide based CO catalysts, NANOCAT® showed a significantly enhanced catalytic performance.

Compared to supported metal catalysts, using iron oxide as a CO catalyst also has other advantages. Iron oxide can be used as a catalyst for CO oxidation in the presence of oxygen, and, in the absence of oxygen, as a direct CO oxidant as well by losing the lattice oxygen. The resulting reduced forms such as Fe could further catalyze the disproportionation reaction of CO, with carbon and CO_2 as products. Iron oxide is inexpensive and readily available, rendering it ideal as a disposable catalyst for one-time uses.

The application of a catalyst in the pyrolysis of biomass to remove CO might also change the distribution of other products and thus reduce the yields of desirable ones. For example, the pyrolysis of biomass produces numerous specific compounds ranging from hydroxyacetealdehyde and acetic acid to levoglucosan,[6] and the relative abundance of these compounds could be changed with the application of CO catalysts. Therefore, it is necessary to study the effects of NANOCAT® on the product distribution in the pyrolysis of biomass and to extract the chemical information on the environment of thermal processing, which can then be used to adjust the process parameters more precisely.

This work begins with the characterization of the structure of starting NANOCAT®. Zhao *et al* in their 1993 publication[2] concluded, based largely on EXAFS data, that NANOCAT® consisted of nanoparticle lepidocrocite (FeOOH). Because of the time-lag between their study and our experimentation, as well as confusion about the possible phases present, it was decided to re-examine the structure of the starting material as well as to examine the structures of products after the experiments. Secondly, the experimental data of NANOCAT® obtained from a flow tube reactor is summarized according to its three utilities: (1) as a catalyst for CO oxidation by oxygen; (2) as an oxidant by itself for CO oxidation; and (3) its reduced forms (Fe, *etc.*) as a catalyst for the CO disproportionation reaction. A more detailed kinetic study of the catalytic oxidation of CO on NANOCAT® can be found in the literature.[5] The superior performance of NANOCAT® as a CO catalyst is linked to its small particle size and the existence of a FeOOH phase. Thirdly, the catalytic and oxidative effects of NANOCAT® on other pyrolytic products of biomass compounds such as cellulose (Avicel), vanillin and stigmasterol were examined by Molecular Beam Mass Spectrometry (MBMS), which allowed direct and real time sampling from the pyrolyzing system. Multivariate factor analysis was employed to simplify the data and discover underlying changes in chemistry that may not be obvious from the direct comparisons of the complex mass spectra obtained under different conditions. The phase changes of the NANOCAT® during the reactions under various conditions are also examined and discussed.

24.2. EXPERIMENTAL

24.2.1. Structural Characterization

The materials were studied by high-resolution transmission electron microscopy (HRTEM) using an FEI Tecnai F20 transmission electron microscope. The microscope, equipped with a field emission gun, was operated at 200kV accelerating voltage. The images were collected digitally using a Gatan imaging filter (GIF). These images were then transformed by a fast Fourier transform using the Gatan Digital Micrograph software. Interplanar spacings and interplanar angles were measured directly from the transformed images. Selected area electron diffraction patterns (SAED) were collected by a video camera mounted within the 35 mm port of the microscope. Electron energy loss spectra (EELS) were also collected with the GIF. Energy dispersive spectra (EDS) were collected with an EDAX spectrometer and analyzed by Emispec's Tecnai Image Analysis (TIA) software. Fourier transform infrared spectroscopy (FTIR) was taken with a Nicolet FTIR instrument using a diamond cell. Powder X-ray diffraction patterns were taken with a Philips X'pert X-ray diffractometer. XAFS data were collected by the Naval Research Laboratory (NRL, Washington DC.), using the synchrotron beam facility in the Brookhaven National Laboratory (BNL, Upton, NY.)

24.2.2. Setup for Catalytic and Oxidative Reactions of CO

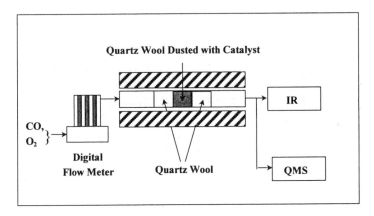

FIGURE 24.1. Schematic diagram of flow tube reactor system.

The characterization of NANOCAT® as a CO catalyst and oxidant was carried out in a flow tube reactor system. A schematic diagram of the system is shown in Figure 24.1. A piece of quartz wool dusted with a known amount of NANOCAT® was placed in the middle of the flow tube (length: 50 cm, I.D: 0.9 cm), sandwiched by the two other pieces of quartz wool. The quartz flow tube was then placed inside a Thermcraft furnace controlled by a temperature programmer. The sample temperature was monitored by an Omega K-type thermocouple inserted into the dusted quartz wool. Another thermocouple was placed in the middle of the furnace, outside the flow tube, to monitor and record the furnace temperature. The temperature data were recorded by a Labview-based program written in-house. The temperatures of the catalyst bed are used in all plots and kinetic calculations. The inlet gases were controlled by a Hastings digital flow meter. The gases were mixed before entering the flow tube. The effluent gas was analyzed either by an NGA2000-MLT multi-gas analyzer manufactured by Rosemount Analytical (non-dispersive near infrared detectors for CO and CO_2, paramagnetic detector for oxygen), or a Balzer Thermal Star Quadrupole Mass Spectrometer (QMS) through a sampling capillary. When the mass spectrometer was used as the monitor, a 15% contribution from the fragmentation of CO_2 (m/e=44) to CO (m/e=28) was accounted for.

24.2.3. MBMS Reactor System for Pyrolysis of Biomass Model Compounds

The effects of NANOCAT® on pyrolytic products of biomass compounds were examined by molecular beam mass spectrometry (MBMS). All reactions were carried out under atmospheric pressure in a quartz tube reactor coupled to a MBMS for product detection. The MBMS was developed by the National Renewable Energy Laboratory (NREL, Golden, CO.) and fabricated by TDA Research Inc. (Golden, CO.). The quartz reactor consisted of inner and outer tubes, with pyrolysis occurring in the inner tube, which has a 5 mm i.d. and a 40 cm length. An electric furnace was set with two temperature zones. One was at 450 °C for pyrolysis and the other at 350 °C for catalysis as shown in Figure 24.2. The temperature at the second zone was varied to study its effect

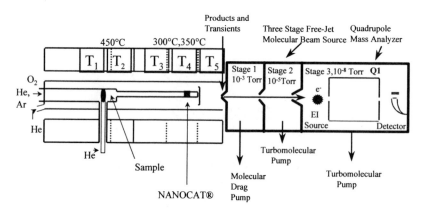

FIGURE 24.2. Schematic of tubular gas phase pyrolysis reactor coupled to a molecular-beam mass spectrometer sampling system.

on catalysis. The flow rate of carrier gas (He) was 500 ml/min. in the inner tube, keeping the total flow rate (inner and outer flow) at 10,000 ml/min. to satisfy the demands of the stage 1 orifice. 10 ml/min. of argon was used as an internal standard to get semi-quantitative data. To study the effect of oxygen on the catalysis, 3% of oxygen was introduced into the reactor for some experiments. A catalyst bed was located at one end of the inner tube with 2.5 mg of NANOCAT® placed between two pieces of clean quartz wool. The catalyst bed was preheated to reach the designated temperatures before reactions were carried out. A biomass sample (20 mg) contained in a quartz holder, or "boat", was inserted into flowing, preheated helium carrier gas. The carrier gas was introduced through the ends of both inner and outer tubes. Vapors formed from the pyrolysis zone passed through the catalyst bed and the final products exiting the reactor (inner tube) were diluted with the helium carrier gas in the outer tube and flowed over the sampling orifice, expanded on the apex of the sampling cone into a low-pressure chamber. The pressure difference was sufficient for free-jet expansion, which quenches the products and allows light gases, high-molecular-weight compounds, and reaction intermediates to be simultaneously sampled and analyzed. A molecular beam, collimated through a second expansion, entered the ion source of single quadrupole mass spectrometer (Extrel), with axial geometry, which did not disrupt the molecular beam. Low-energy crossed beam electron ionization (approximately 25 eV) was used to minimize fragmentation within the ion source. The mass range of interest, m/z 15 to m/z 500, was repeatedly scanned and the data cumulatively stored over the course of the pyrolysis experiments.

There were four different reaction conditions studied: (1) pure pyrolysis (labeled as "pyrolysis" in the figures); (2) pyrolysis with 3% O_2 in the system (as "oxidation"); (3) pyrolysis with 3% O_2 and NANOCAT® (as "catalysis"); and (4) pyrolysis with NANOCAT® but no O_2 (as "py-catalysis").

Multivariate analysis was used to reduce the data by finding correlated masses that can be expressed by new variables, the *factors*, and determine groups of products present in the mass spectra. The software package used was the ISMA (Interactive Self-modeling Multivariate Analysis) program.[7]

24.2.4. Materials

The NANOCAT® sample was purchased from MACH I Inc. (King of Prussia, PA.). The average particle size was 3 nm, according to the manufacturer, and the sample was used without further treatment. The FeOOH (goethite) was purchased from Aldrich and the original particle size was reported to be between 30 to 50 μm. It was ground to an average diameter of 4.5 μm before use. The α-Fe_2O_3 was also purchased from Aldrich with an average particle size of 5 μm and it was used without further treatment. The CO (4%), and O_2 (21%) gases, all balanced with Helium, and mixtures of 3.44% CO with 20.6% O_2, also balanced with helium, were purchased from BOC Gases with certified analysis.

The model biomass compounds of cellulose (Avicel), vanillin, and stigmasterol were purchased from FCI (>99% pure), Fisher Scientific (>99% pure), and Aldrich (>95% pure), respectively, and used as received.

24.3. RESULTS AND DISCUSSION

24.3.1. Structure of NANOCAT®

NANOCAT® is a brown colored, free flowing nanoparticle powder with a bulk density of only 0.05 g/cm³. HRTEM showed the majority of the NANOCAT® particles to be approximately 3 nm in diameter, as claimed by the manufacturer. For the most part, the grains were anhedral and HRTEM images indicated a very distorted lattice as shown in Figure 24.3. Fourier transforms of portions of the images were useful to obtain average d-spacings. Few large d-spacings were measured, however, and cross fringes were generally rare. Phase determination was attempted by compiling a list of the observed d-spacings and searching the ICDD (International Center for Diffraction Data)[8] database for suitable matches. The closest matches by this method were bernalite ($Fe(OH)_3$) and lepidocrocite (FeOOH). However, the largest d-spacings associated with these phases were not observed.

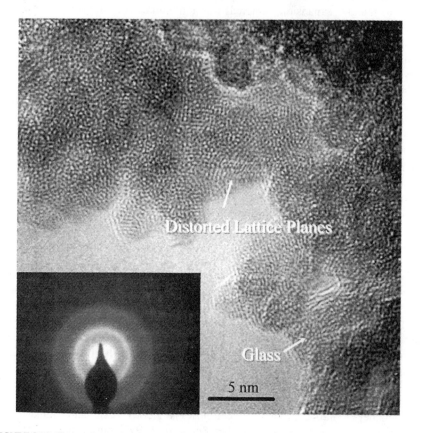

FIGURE 24.3. The starting material possessed a weak, highly distorted lattice as well as glassy material.

Some of the grains had no apparent crystal structure at all, and were apparently oxide glasses. An example can also be seen in Figure 24.3. All of these grains had approximately 3 nm diameters, as mentioned above. In addition, a few larger grains were also observed in the sample, ranging in size from 3 nm to 100 nm. These grains had a much better organized structure with the appearance of larger d-spacings and could be readily indexed. The d-spacings of these larger grains were consistent with γ-Fe_2O_3 (maghemite).

EDS analyses performed in HRTEM suffer from a number of effects that can adversely affect confidence in quantification. These include unknown sample thicknesses, geometries, and non-linearity of low energy responses of the detector for light elements. As a result, oxygen analysis is usually qualitative, rather than quantitative. In order to circumvent, at least partially, these limitations, machine parameters were adjusted in the quantification software to achieve reasonable results from oxide compositions of known materials, including aluminum oxide, silicon dioxide, and iron oxides. When energy dispersive spectra were quantified with these adjusted parameters, the results consistently showed NANOCAT® to be oxygen enriched, relative to Fe_2O_3. In fact, the oxygen/iron atomic ratio ranged from 1.7 to 2.8 with an average of 2.23. This compared nicely to the ratio for goethite (and lepidocrocite) and bernalite of 2 and 3, respectively, suggesting that the excess oxygen might be a result of hydroxide within the crystal lattices.

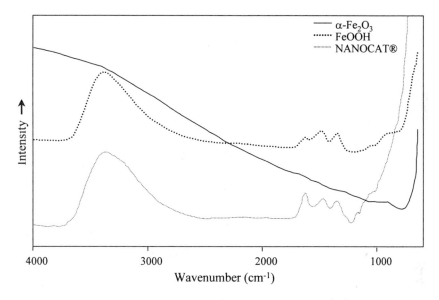

FIGURE 24.4. Infrared Spectra of some iron oxides. The O-H stretching mode had a much larger intensity for NANOCAT® and FeOOH than for α-Fe_2O_3.

Infrared spectra of NANOCAT®, ground FeOOH (goethite), and α-Fe_2O_3 were taken. All three materials are powder samples, with NANOCAT® having the smallest grain-size diameters. The spectra were collected in an FTIR spectrometer in a diamond cell and the background uncorrected spectra are shown in Figure 24.4. The group of small adsorption bands in 1300-1700 cm^{-1} region are associated with the bending modes of hydroxyl and the large absorption band appearing at about 3400 cm^{-1} is normally attributed to the hydroxyl stretching mode. This band may be observed in spectra of NANOCAT® and FeOOH, but is conspicuously absent from that of Fe_2O_3. The breadth of the band suggested either that the environment of the O-H bond within the solid is highly variable, or that non-bonded H_2O may have contributed to the absorption. Considering the small size of these materials, neither scenario is unlikely. While it is possible that much or all of the hydroxyl stretch observed resulted from adsorbed water, the strong similarity of NANOCAT® to the FeOOH spectrum and the near absence of this band in Fe_2O_3 suggested that the hydroxyl stretch is largely the result of lattice hydroxyl and not adsorbed water. A small shoulder at about 3200 cm^{-1} in both NANOCAT® and FeOOH spectra may represent the absorption band of well-crystallized, hydroxylated NANOCAT® and goethite phases respectively. A band at a slightly lower frequency than that of the water band is not unusual for other hydroxylated minerals, such as apatite.

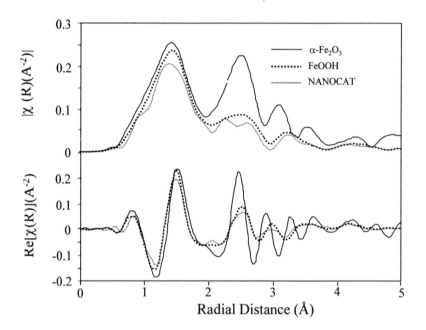

FIGURE 24.5. Fourier transform of $k\lambda(k)$ at Fe K-edge. The top panel shows the magnitude of the complex Fourier transform and its real part is shown in the bottom panel. The transforms were performed on $\lambda(k)$ data with $k_{min} = 2Å^{-1}$ and $k_{max}=12Å^{-1}$.

These samples were also sent to Naval Research Laboratory for EXAFS analysis. The results are shown in Figure 24.5. The Fourier transform of the real part of the function is shown to emphasize the frequencies contributing to the spectra. The NANOCAT® sample displayed a marked similarity to that of FeOOH yet was quite different from that of Fe_2O_3.

EELS, like EXAFS, can, at least in theory, provide information on composition, bonding characteristics, and coordination. EELS spectra were taken of NANOCAT®, Fe_2O_3, FeOOH, and Fe_3O_4. In the low loss region, a broad, but unusually large intensity appeared at about 12 eV for the NANOCAT® and FeOOH samples. This is roughly the same as the position of the hydrogen K edge. By contrast, both Fe_2O_3 and Fe_3O_4 have a considerably reduced band in that position. In summary, EDS spectra, EXAFS spectra, IR spectra, and EELS all suggested an oxy-hydroxide component to NANOCAT®.

24.3.2. NANOCAT® as a Catalyst and Oxidant for CO Oxidation

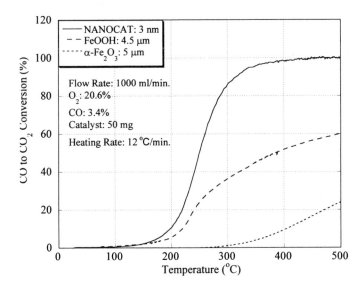

FIGURE 24.6. Comparison of NANOCAT® vs. FeOOH vs. α-Fe$_2$O$_3$ in catalytic oxidation of CO.

Iron oxide, like several other transition metal oxides such as manganese oxide and nickel oxide, has been known to be a catalyst for CO oxidation for some time [4]. However, the physical and structural characteristics of NANOCAT®, as discussed in the previous section, are quite different from those of other micron size iron oxides. It is reasonable to expect that the very small particle size of 3 nm might contribute to an improvement in the catalytic performance of NANOCAT®. NANOCAT® was evaluated as a catalyst for CO oxidation in a flow tube reactor, along with 5 μm α-Fe$_2$O$_3$ and FeOOH powder for comparison. The tests were conducted under identical conditions. Because the bulk density of NANOCAT® was much smaller than that of α-Fe$_2$O$_3$ and FeOOH, all three samples were dusted onto the same amount of quartz wool so that the volumes (as well as mass) of the catalysts in the flow tube were the same, thereby fixing residence times to be the same as well. As shown in Figure 24.6, the catalytic performance of NANOCAT® was significantly better than the 5 μm size α-Fe$_2$O$_3$ powder. 50 mg of the NANOCAT® could catalyze the oxidation of more than 95% of the CO to CO$_2$ at 350 °C with an inlet gas mixture of 3.44% CO and 20.6 % O$_2$ at the total flow rate of 1000 ml/min. Under identical conditions, the same amount of the α-Fe$_2$O$_3$ powder, with a particle size of 5 μm, could only catalyze oxidation of about 5% of the CO to CO$_2$. In addition to that, the initial light off temperature for NANOCAT® was at least more than 100 °C lower than that of α-Fe$_2$O$_3$ powder. The fact that the small mass of NANOCAT® could provide such a high CO conversion efficiency can probably be attributed, at least partially, to its small particle size of 3 nm and high surface area of 250 m^2/g. However, the high surface area alone cannot explain why the onset of the activation temperature of NANOCAT® was at least 100 °C lower than the onset temperatures of iron oxide powder. The CO conversion trace of FeOOH, also shown in Figure 24.6, can shed some light into this matter. The particle size of FeOOH powder was 4.5 μm, which is very close to the particle size of 5 μm α-Fe₂O₃ and much larger than the particle size of NANOCAT®. However, the onset temperature of FeOOH as a CO catalyst was identical to that of the NANOCAT®, although the CO conversion at 350 °C was less than 50%. Therefore, it is reasonable to conclude at this point that for NANOCAT®, the small particle size and large surface area of the powder contributed to the high CO conversion efficiency and the FeOOH component contributed to the low onset temperature.

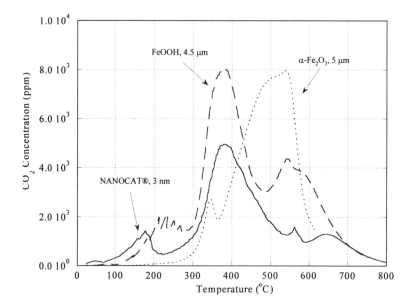

FIGURE 24.7. Comparison of NANOCAT vs. FeOOH vs. α-Fe₂O₃ in reduction by CO. Reprinted from Figure 6 in reference [5], Copyright (2003), with permission from Elsevier.

The reason that FeOOH is more active than Fe₂O₃ at low temperature can be explained as follows. The catalytic oxidation of CO by metal oxides may be generally broken down into two steps:

$$MO + CO = M + CO_2 \qquad (1)$$

$$M + \tfrac{1}{2} O_2 = MO \qquad (2)$$

where M is the metal. The first step of the catalytic reaction is for MO to lose one oxygen atom to CO to form CO₂. The tendency of MO to lose one oxygen atom to CO should be an important factor in determining how good a catalyst MO is. The MO reaction with CO in the absence of oxygen can be checked experimentally. Figure 24.7

shows the results of experiments with NANOCAT®, α-Fe_2O_3, and FeOOH powder following Eq. (1). In Figure 24.7, CO_2 production is used as the indicator of the reaction progress. It can be seen that despite the vastly different particle sizes (3 nm vs. 4.5 µm) the traces of NANOCAT® and FeOOH are remarkably similar and the onset temperatures are very close. On the other hand, the onset temperature for α-Fe_2O_3 is significantly higher and the general shape of curve is quite different as well. The onset temperatures for NANOCAT®, α-Fe_2O_3, and FeOOH in Figure 24.7 mimic the onset temperatures shown in Figure 24.6. The result seems to indicate that it is easier for FeOOH to lose lattice oxygen from Fe-O bonds than α-Fe_2O_3. One probable reason may be because FeOOH has a slightly longer Fe-O bondlength and thus a slightly weaker Fe-O bond. The average Fe-O bondlength in α-FeOOH is reported to be 2.02 Å by neutron diffraction measurement.[9] In α-Fe_2O_3, the value is 1.99 Å.[10] Lattice distortion resulting from dehydration could also emphasize this difference. This set of tests suggests that the FeOOH component present in NANOCAT® provides the catalytic activity at low temperatures and the small particle size provides the high conversion efficiency.

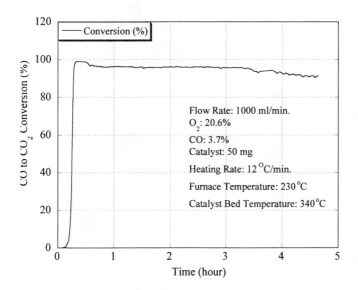

FIGURE 24.8. Test of thermal stability of NANOCAT®.

FeOOH will proceed with dehydration upon heating as follows:

$$2FeOOH = Fe_2O_3 + H_2O \qquad (3)$$

It is therefore important to check the thermal stability of the phases present in NANOCAT® during the catalytic process. A test was carried out by increasing and then holding the catalyst's temperature at 340 °C while the CO conversion was monitored constantly, as shown in Figure 24.8. It can be seen that the CO conversion was stable at above 90% for more than four hours. The thermal stability with respect to catalytic activity of NANOCAT® was quite satisfactory, at least for one-time use application such as in cigarettes.

Heating of NANOCAT® alters the structure of the material and it would be expected that upon heating, the material would transform to an anhydrous phase consistent with classical thermodynamics. In the presence of 3% O_2 that phase would be α-Fe_2O_3. Figure 24.9 shows the results of a heating experiment at 350 °C. Selected area electron diffraction shows patterns consistent with the rhombohedral α-Fe_2O_3 phase. In addition, the heated material had coarsened to produce a grain size of 15 nm or greater. The new crystals filled the void spaces between the original 3 nm particles and hence do not represent annealing. Instead, this is reminiscent of classical crystal growth in the presence of a fluid phase acting as a flux. The only fluids present are the gases and evolved water, suggesting that the flux may have been the waters of dehydration of the original FeOOH component.

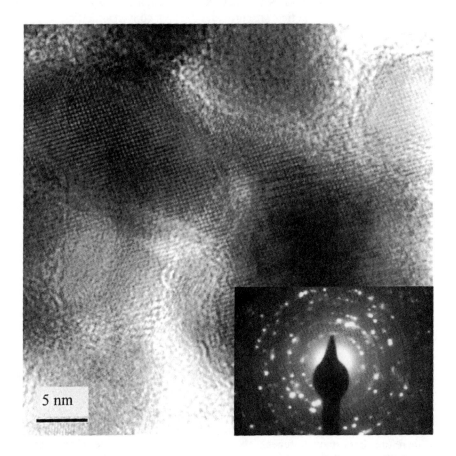

5 nm

FIGURE 24.9. Sample heated to 350 °C with oxygen. The solid phases were entirely α-Fe_2O_3. Inset is an SAED (selected area electron diffraction patterns).

The measurements of apparent activation energy and the reaction orders for CO and O_2 were described in details elsewhere.[5] Here only a brief description of the test method and results is given. The reaction order for CO was measured isothermally at 244 °C, at which temperature the CO to CO_2 conversion percentage was about 50%. In the inlet gas, the CO concentration was varied from 0.5 to 2.1% while the O_2 concentration was maintained at 11%. It was found that the catalytic oxidation of CO on NANOCAT® was first order with respect to CO. The reaction order of O_2 was measured in a similar fashion. Care was taken to make sure that the O_2 concentration was not lower than ½ of the CO inlet concentration, based on the stoichiometry of the reaction. The purpose was to prevent any direct oxidation of CO by NANOCAT® because of insufficient O_2. In the inlet gas the O_2 concentration was varied from 0.4 to 1.6% while the CO concentration was maintained at 0.79%. The increase of the O_2 concentration had very little effect on the CO_2 production. Therefore, it can be concluded that the reaction order of O_2 was approximately zero. Walker et al[4] also observed a first order reaction for the overall catalytic CO reaction on 100 mesh Fe_2O_3/TiO_2 catalyst. Since the overall reaction was a first order reaction, the apparent activation energy E_a can simply be measured by fitting CO conversion traces such as those shown in Figure 24.6 with the Arrhenius equation. The average measured E_a for NANOCAT® was found to be 14.5 kcal/mol, which is considerably lower than other Fe_2O_3 based catalyst ($\approx$ 20 kcal/mol).[4]

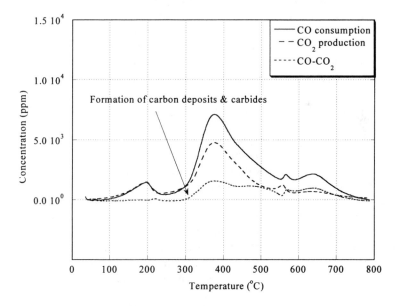

FIGURE 24.10. Reduction of NANOCAT® by CO and observation of CO disproportionation. Reprinted from Figure 7(B) in reference [5], Copyright (2003), with permission from Elsevier.

In the absence of O_2, Fe_2O_3 can also behave as a reagent to oxidize CO to CO_2 with sequential reduction of the Fe_2O_3 to produce reduced phases such as Fe_3O_4, FeO and Fe.[11] The ability to provide lattice oxygen is important in certain potential applications, such as in a burning cigarette, where the amount of O_2 present is insufficient to oxidize all the CO formed. In such instances, Fe_2O_3 can function as a catalyst first, and then function again as an oxidant. In this way, a maximum amount of CO can be converted to CO_2 with only a minimal amount of Fe_2O_3 added to the system.

The reaction of NANOCAT® with CO in the absence of oxygen is quite complicated. First, the Fe_2O_3 will be reduced stepwise to Fe_3O_4, FeO, and Fe, as the temperature increases, according to the net equation of

$$Fe_2O_3 + 3CO = 2Fe + 3CO_2 \qquad (4)$$

The freshly formed Fe can then catalyze the disproportionation reaction of CO,[12-14] producing CO_2 and a carbon deposit,

$$2CO = C + CO_2 \qquad (5)$$

The carbon can also react with the Fe to form different species of iron carbides, such as Fe_3C or Fe_3C_7, and thus deactivate the Fe catalyst. However, these carbides could also be involved in the disproportionation reaction. Eventually, when the surface is completely covered by carbon deposits, the disproportionation reaction of CO stops.

For the direct oxidation experiment, only 4% CO balanced by helium was used in the gas inlet. The CO and CO_2 concentrations were monitored in the gas outlet while the temperature was increased linearly from ambient to 800 °C at the rate of 12 °C per minute. However, the production of CO_2 does not account for all of the depletion of CO, as shown in Figure 24.10. There is more CO depleted than CO_2 produced. The difference between the CO depletion and the CO_2 production, as indicated by the dashed line, starts to appear at 300 °C and extends all the way to 800 °C. In the disproportionation reaction of CO catalyzed by Fe, the CO consumed is twice as much as the CO_2 produced, and there should be carbon deposited on the surfaces and/or the formation of iron carbide on the surfaces.

To further confirm that a disproportionation reaction occurred in the presence of CO, the NANOCAT® sample was heated in a He/CO atmosphere for several hours. At 300 °C the material produced was black in color. HRTEM analyses indicated the presence of large, 40 nm grains (Figure 24.11). EDS analyses were somewhat variable and indicated the presence of carbon, iron, and a small amount of oxygen. Fourier transforms of some of the grains were consistent with iron carbides. The oxygen may have been from oxides (perhaps wüstite) entrained within these carbides and the carbide grains were also coated with carbonaceous material. The carbon phase was not graphite, but consisted of graphene sheets with variable spacings between the sheets. Nevertheless, the nearly parallel alignment of the sheets on the surfaces of the carbide grains was reminiscent of graphite.

Increasing the temperature to 800 °C produced a black product. This material consisted of large, 50-100 nm grains of iron carbide (Figure 24.12) which in turn were coated with a thick, typically 5-10 nm shell of carbon. The carbon was highly crystalline graphite as indicated by the 3.4 Å interplanar (00.1) spacing. Furthermore, cross-fringes indicated that the 3-dimensional structure of graphite was present, rather than just the 2-dimensional structure of graphene sheets. In addition, closer examination of a piece of solid residue found at the bottom of the catalyst bed revealed the presence of metallic iron. These iron grains did not have thick shells of graphite surrounding them.

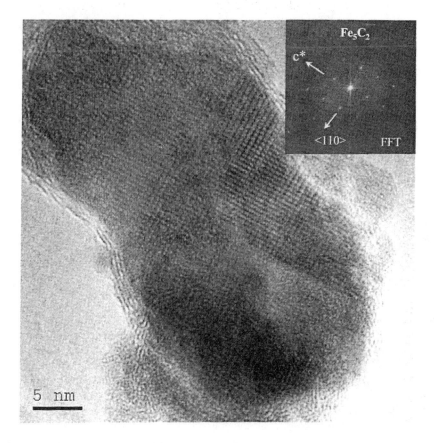

FIGURE 24.11. NANOCAT® heated to 300 °C under a CO atmosphere formed a mixture of different iron carbides.

In the test shown in Figure 24.10, 0.344 mmol of NANOCAT® was used in the experiment. The measured CO consumption by direct oxidation (Eq. (4)) was 1.027 mmol, very close to the stoichiometric value of $3 \times 0.344 = 1.032$ mmol. The measured CO consumption by the disproportionation reaction (Eq. (5)) was 1.048 mmol. Therefore, before the freshly produced Fe catalyst was deactivated by forming iron carbide, it can catalyze a substantial amount of CO to undergo the disproportionation reaction.

In summary, NANOCAT® can be used to remove CO in three ways: catalytic oxidation by O_2, direct oxidation by donating lattice oxygen, and disproportionation by its reduced forms of Fe or carbide. It is especially suitable for some one-time use applications of CO removal.

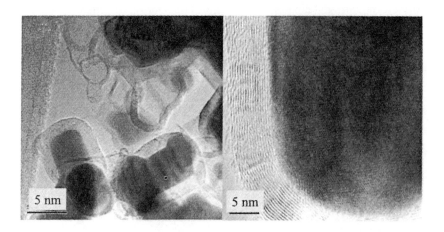

FIGURE 24.12. NANOCAT® heated to 800 °C in a CO atmosphere formed carbides with graphite shells.

24.3.3. The Effects of NANOCAT® on the Pyrolysis of Biomass Model Compounds

Biomass model compounds, cellulose, vanillin and stigmasterol, were selected to develop an understanding of the effects of NANOCAT® on the chemical and physical process of biopolymer pyrolysis. Avicel is a pure cellulose that was used in these experiments. Cellulose is the structural basis of plant cells and consequently is the important constituent in most plant biomass.[15, 16] Vanillin is a model compound for lignin, which is another important biomass constituent.[17] It contains a number of the functional groups found in lignin and it is a significant product from lignin pyrolysis. Stigmasterol is a plant sterol found in many biomass systems. Pyrolysis studies of stigmasterol showed a complex product slate of hydrocarbons due to its condensed structure[18] and therefore it was considered a good candidate for the study of the effects of NANOCAT® on the product distribution of sterol pyrolysis. A detailed approach to the study of biomass pyrolysis is illustrated in previous works.[15-18]

For this study, NANOCAT® was preheated to the reaction temperature before any pyrolysis experiment was conducted. In section 3.1, it was discussed that the phases of NANOCAT® before preheating consisted mostly of iron oxy-hydroxides with a small particle size (3 nm). However, after preheating, NANOCAT® loses water and coarsens. According to electron diffraction patterns, the resulting phase turned out to be α-Fe_2O_3. This is the starting phase of NANOCAT® in the pyrolysis of biomass model compounds. Pyrolysis of Avicel can produce a complex mixture of compounds. For the purposes of comparison, thermo-chemical data from the pyrolysis of Avicel, as detected by the MBMS at 500 °C and 650 °C with 0.6 seconds gas phase residence time, are given in Figure 24.13 where the average mass spectra of the products are shown. Major masses produced at the low severity condition of Figure 24.13 (a) are found at m/z 57, 60, 70, 73, 98, and 144. These are fragment ions of levoglucosan (m/z 162), which is a primary product from the pyrolysis of Avicel.[15, 16] At more severe condition (Figure 24.13 (b)), carbonyl compounds and some of the related hydroxyl derivatives, such as acetaldehyde (m/z 44), acrolein (m/z 56), furan (m/z 68), and furfural (m/z 96) were found. These secondary products were resulted from the further decomposition of the primary pyrolysis product. These prominent masses were selected for the study of their variation as a function of reaction conditions. Additionally, masses at m/z 28 (CO) and 44 (CO_2) were also chosen because of their noticeable changes with the reaction conditions. The product distributions at 350 °C under three different reaction conditions are shown in Figure 24.14. The first data set is from the pyrolysis condition (pyrolysis), *i.e.* in the absence of O_2 (bars with hatch mark). Adding 3% O_2 to the carrier gas (oxidation) did not alter the major product distributions from levoglucosan significantly (bars with vertical lines). However, in the presence of O_2 and NANOCAT® (catalysis), it is obvious that the formation of CO and CO_2 were certainly enhanced at the expense of the major products (solid bars). CO_2 may also contribute to the growth of CO as a fragment ion through the ionization process.

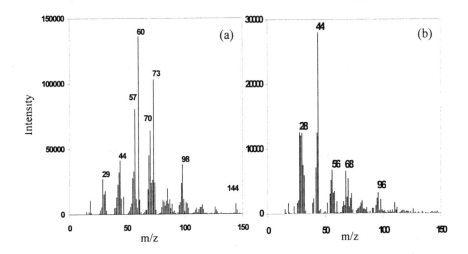

FIGURE 24.13. MBMS spectra (with background correction) of Avicel pyrolyis at (a) 500 °C and (b) 650 °C (residence time = 0.6 sec.).

The effect of temperature on catalysis by NANOCAT® is shown in Figure 24.15, where product distributions at two different temperatures, 300 °C and 350 °C, were compared in the presence of 3% O_2 and NANOCAT®. NANOCAT® served to extend conversion at the elevated temperature of 350 °C, by promoting the formation of CO_2 from the major products, as indicated by the increased intensity of m/z 44 and the decreased intensities of larger masses. In Figure 24.16, two sets of data were compared for the reaction of Avicel at 350 °C in the presence of NANOCAT®, with 3% O_2 (catalysis) and without O_2 (py-catalysis) in the carrier gas. It is can be seen that the presence of O_2 in the carrier gas resulted in the formation of more CO_2 by converting the major products.

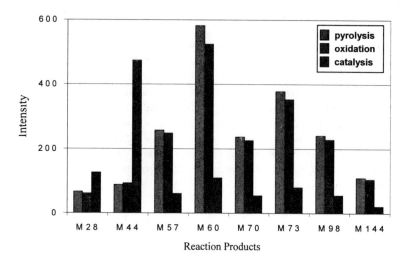

FIGURE 24.14. The variation of Avicel pyrolysis products at 350 °C under three different conditions: pyrolysis, oxidation, and catalysis.

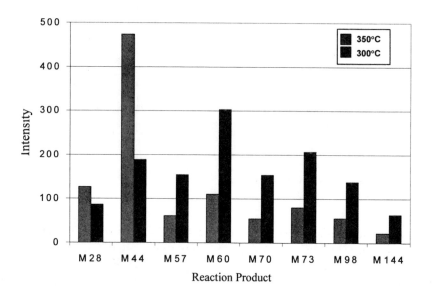

FIGURE 24.15. The effect of temperature on the catalytic pyrolysis of Avicel with 3% O_2.

HRTEM analysis indicated phase changes of NANOCAT® depending on the presence of O_2. NANOCAT® that was exposed to vapors from biomass decomposition in the presence of 3% O_2 was still α-Fe_2O_3. In the absence of O_2, some of NANOCAT® was reduced to Fe_3O_4 so that both phases were observed. Consequently, overall oxidative activity was weakened in the absence of O_2.

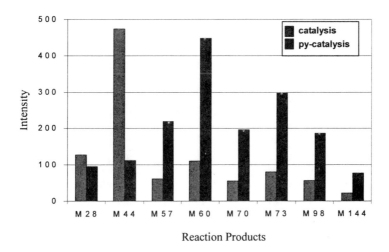

FIGURE 24.16. The effect of oxygen on the pyrolysis of Avicel in presence of NANOCAT® at 350 °C. Catalysis: with O_2 presence; py-catalysis: without O_2 presence.

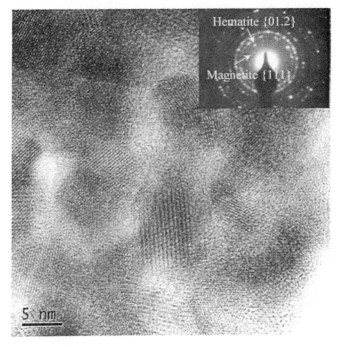

FIGURE 24.17. NANOCAT® heated to 450 °C in the presence of biomass with 3% oxygen. Sample consisted mostly of magnetite Fe_3O_4 and had low catalytic activity. Inset is an SAED.

When NANOCAT® was mixed with the biomass without pre-heating (unlike the former experiments where pre-heating converted it to α-Fe_2O_3 and they were kept separated in the furnace) and heated to 350 °C in the presence of 3% oxygen, both hematite and magnetite were formed. Much less coarsening was observed than when they were not mixed, although considerable variation in grain-size was found. This could simply be a reflection of the variability of the transport of gas to the oxide surfaces within the biomass-catalyst bed. However, many of the particles were still of a 3 nm size, even though they had recrystallized to either magnetite or hematite, suggesting that solid state diffusion was operative in the transformation, producing pseudomorphs of the original material. Heating to 450 °C did not change the results appreciably. Both phases were present although some coarsening was evident, as shown in Figure 24.17. The efficacy of NANOCAT® under these conditions had been considerably reduced, lending evidence that the oxidized phase was the active phase. The magnetite phase is not stable in the presence of 3% O_2 at this temperature and would be expected to convert to an oxidized form. However, since magnetite is a spinel, it may be easier to convert to an oxidized spinel rather than undergo complete rearrangement of the lattice to hematite. If so, one might expect to see distortions of the spinel structure to accommodate non-stoichiometry. Figure 24.18 shows one such distortion that was observed during the course of these experiments. The figure shows planar defects on one of the {100} planes of magnetite.

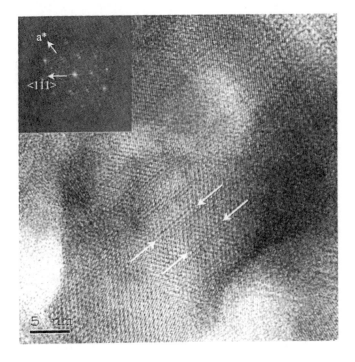

FIGURE 24.18. NANOCAT® heated to 350 °C in the presence of biomass and 3% O_2. Distorted spinels commonly had defects (denoted by arrows in the image). Inset is an SAED indexed for magnetite.

Removal of a set of Fe atoms on this plane could result in a more oxidized state without a massive change in structure. In the absence of oxygen, the grain size of NANOCAT® was essentially unchanged, however, the material did recrystallize as shown in Figure 24.19. Electron diffraction patterns indicated that it had been entirely converted to magnetite. The absence of trace quantities of hematite, or of pseudomorphs of the rhombohedral hematite, suggested that the conversion proceeded directly to magnetite. Again, this may reflect a spinel-like nature of the original starting material.

FIGURE 24.19. NANOCAT® mixed with biomass and heated to 350 °C without oxygen. The phases were spinels and approached the Fe_3O_4 structure. Inset is an SAED.

It is difficult to extract the underlying changes in chemistry directly from the complex raw data obtained from the MBMS. Multivariate factor analysis using ISMA software package[7] reduce the data by finding correlated masses that can be expressed by a new single variable or *factor*. The data reduction makes possible a graphical display of the data that not only shows trends but also provides chemical insight. Each mass has a correlation coefficient, "the loading", with each of the *factors*. The masses with high loadings on a particular *factor* are correlated and that *factor* represents that group of masses and is quantified for each sample by the factor score for a particular sample. By use of the factor loadings for the masses and factor scores for the samples, factor analysis can reflect the presence of chemical components or trends in the data set.

This factor analysis was applied to two time-resolved data sets from the decomposition of Avicel under identical conditions, but one with NANOCAT® and the other without NANOCAT®. Only two significant groups of masses, *factor 1* and *factor 2* were extracted. Figure 24.20 exhibits the variation of the factor score for each *factor*, which reveals time-resolved behavior of *factors* in the absence (Figure 24.20 (a)) and presence of NANOCAT® (Figure 24.20 (b)). It is apparent that the *factors* were completely switched from the data in Figure 24.20 (a) to the data in Figure 24.20 (b).

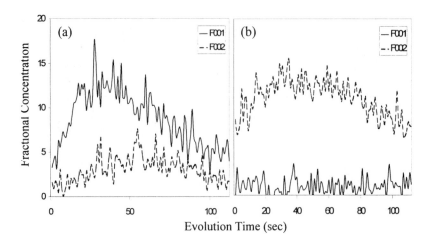

FIGURE 24.20. Time resolved factor score plot of Avicel decomposition in the (a) absence and (b) presence of NANOCAT®, at 350 °C with 3% O_2.

The mass spectra for two *factors* provide insight into the chemistry changes that occurred due to the different reaction conditions. The factor spectra for each lumped product group, or *factor*, are shown in Figure 24.21. The mass spectrum of *factor 1* corresponds with the mass spectrum of levoglucosan, the primary product from Avicel pyrolysis. Major products in the spectrum of *factor 2* are CO_2 and CO. It also includes small peaks at m/z 68 and 96 that were observed at 650 °C from thermo-chemical conversion (Figure 24.13 (b)) by subsequent cracking of the primary products. This factor analysis shows that the predominant products in the absence of NANOCAT® are just the primary products of Avicel pyrolysis. But in the presence of NANOCAT®, most of the primary products convert to CO_2 and CO. These results are consistent with the observations made with raw data. However, the factor analysis data showed clear trends resulting from the presence and absence of NANOCAT® in the reaction.

Figure 24.22 shows the raw mass spectra resulting from vanillin pyrolysis at 500 °C and 650 °C with a 0.6 second gas phase residence time.[17] The spectrum at 500 °C contains features arising from un-decomposed vanillin (molecular weight = 152 amu) while the spectrum at 650 °C pyrolysis clearly shows that vanillin had reacted to a large extent and had been converted to a number of other compounds. We selected vanillin (m/z 152) and its fragment ion (m/z 109), the subsequent pyrolysis products from vanillin (m/z 150, 138, 66, 28) and CO_2 (m/z 44) to study the effect of NANOCAT® on their

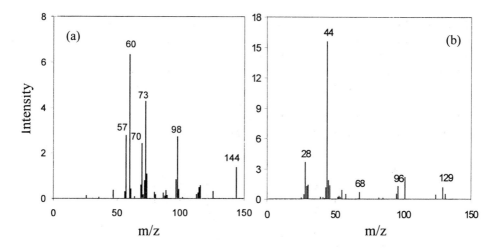

FIGURE 24.21. Component spectra for *factor 1* (a) and *factor 2* (b) of cellulose decomposition at 350 °C with 3% O_2.

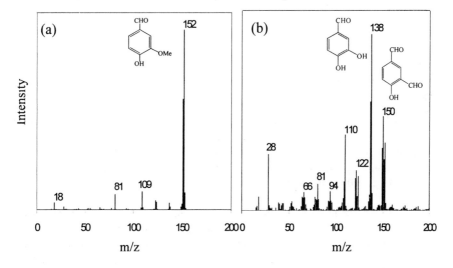

FIGURE 24.22. MBMS spectra (with background correction) of vanillin pyrolysis at (a) 500 °C and (b) 650 °C (residence time = 0.6 sec.). Reprinted from Figure 1 in reference [17], Copyright (2001), with permission from Elsevier.

distribution. Figure 24.23 exhibits the variation of product distribution in the absence and presence of NANOCAT®. NANOCAT® played a significant role in the decomposition of vanillin, not only to convert vanillin to CO_2, but also to produce lower molecular weight compounds with intact aromatic rings (m/z 150, 138) or with an expulsion of CO to form a cyclopentadiene (m/z 66). Unlike Avicel decomposition, where the primary products were converted mostly to CO_2 in the presence of NANOCAT®, vanillin decomposition apparently had a higher energy barrier to be converted to CO_2 due to its structure. Its decomposition produced small molecules.

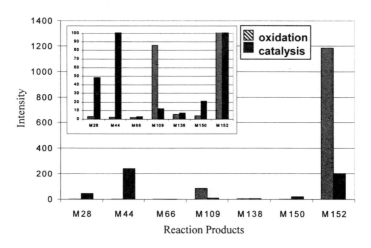

FIGURE 24.23. Effect of NANOCAT® on product distribution of vanillin decomposition at 350 °C with 3% O_2. Insert is the enlarged lower part of the figure.

The MBMS data from the thermo-chemical conversion of stigmasterol under helium is given in Figure 24.24, where spectra at 500 °C and 650 °C with a 0.6 second gas phase residence time are presented.[18] The mass spectrum for the lower reaction severity of 500 °C includes a peak at m/z 412, the molecular weight of stigmasterol, as shown in Figure 24.24 (a); other major peaks in this spectrum are odd masses and are probably fragment ions.

The mass spectrum from the higher reaction severity of 650 °C presents a large number of pyrolysis products, as shown in Figure 24.24 (b). Representative masses from these two spectra were chosen to compare the products distribution of stigmasterol conversion under two new reaction conditions, one without NANOCAT® and the other with NANOCAT® as shown in Figure 24.25. Like two above cases of Avicel and vanillin, the experiments were carried out at 350 °C with 3% O_2 in the carrier gas. The conversion of stigmasterol for both non-catalytic and catalytic conditions is low at this temperature; nevertheless, the catalytic effects on the products distribution of stigmasterol decomposition coincided with those observed in the decomposition of vanillin, where the growth of the lower molecular weight compounds occurred along with an increase in the formation of CO_2 in the presence of the NANOCAT®.

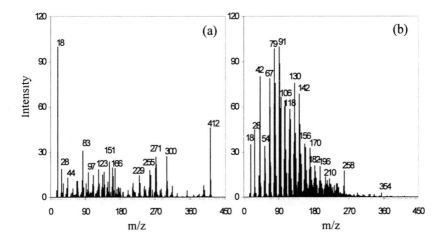

FIGURE 24.24. MBMS spectra (with background correction) of stigmasterol pyrolysis at (a) 500 °C and (b) at 650 °C (residence time = 0.6 sec.). Reprinted from Figure 1 in reference [18], Copyright (2001), with permission from Elsevier.

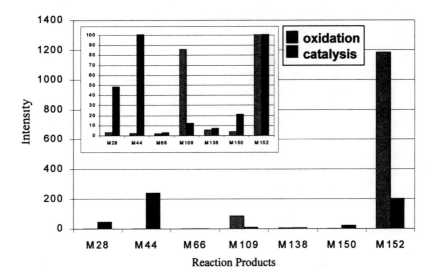

FIGURE 24.25. Effect of NANOCAT® on product distribution of stigmasterol decomposition at 350 °C with 3% O_2. Insert is the enlarged lower part of the figure.

24.3.4. Structural Changes of NANOCAT® during Reactions-A Further Discussion

NANOCAT® is reported to be made by a plasma process, according to its manufacturer. As such, it may be expected that such a rapid crystallization process might have considerable variability in the state of the crystallinity produced. The variability observed in high-resolution images, from glassy to a well-defined spinel, would bear this out. In addition, the presence of water or hydroxyl is evidenced by IR, EELS, EDS, and EXAFS. The broad signals observed in EELS and FTIR would attest to the variability of the structural environment about the hydroxyls. This may also manifest itself by the variability in composition, as was observed in the EDS spectra. Certainly, the glassy grains should have less control over their composition than in a well-defined crystalline lattice. HRTEM indicated small glassy grains, small crystalline grains with large d-spacings missing, and larger grains with the maghemite structure. This may be a reflection of the degree of recrystallization of the solid, tending toward a spinel structure. One can envision a "proto-crystal" as having a cubic anion lattice, essentially like that of a spinel, but with some of the oxygen being hydroxylated and the iron randomly distributed among the tetrahedral and octahedral sites. The average structure of such a material would not have an especially long-range order resulting in a lack of observed large d-spacings. Glide planes and screw axes would also be missing in such a random structure. With increasing temperature or time, dehydration of the hydroxyls and minor diffusion of the iron would eventually order the metal sites and give rise to the larger d-spacings of maghemite or magnetite depending on the valence state of the iron and ordering of the sites to yield the diamond glide and face centering found in the magnetite structure. However, maghemite is not stable relative to other phases. The ultimately stable phase depends on oxygen fugacity. In a highly oxidizing environment, maghemite should convert to α-Fe_2O_3. This was precisely what was observed when the material was heated to 350 °C with 3% O_2. Similarly, it should convert to magnetite if the oxygen concentration was low enough and indeed, this too was observed.

This simple scenario is complicated by the addition of exposure to biomass pyrolysis. When the catalyst resides downstream from the pyrolyzing biomass it's valence was largely governed by the inlet gas concentration. Therefore, in the experiments performed in 3% O_2, only hematite (αFe_2O_3) was formed. However, the hematite filled void spaces between the original particles rather than produced pseudomorphs of the starting material. The material was thus a cement rather than a sintered or annealed product and solid state diffusion was not the operative mechanism for this phase transformation. Instead, mass transport by a fluid phase was suggested. Similarly, in the absence of oxygen, the large cement crystals were magnetite. Solid state diffusion was again not the dominant mechanism. The phase, however, was governed by the inlet gas composition and the extent to which residual α-Fe_2O_3 could buffer the oxygen tension.

By contrast, experiments performed by mixing the biomass with NANOCAT® did show pseudomorphs of the original starting material. Thus, in these cases solid state diffusion was an important operative mechanism. It was somewhat surprising that magnetite formed in the presence of 3% O_2. Apparently, the pyrolyzing biomass was sufficiently reducing to locally diminish the oxygen tension to allow stability of this phase. Reduction would have been accomplished during the catalysis process but without concomitant oxygen to regenerate the catalyst.

The reduced efficacy of the catalyst in the oxidizing experiments when magnetite was present suggested that this phase is not very catalytically active. Furthermore, the oxygen produced through the reduction of the iron oxide was insufficient to transform much of the CO to CO_2. At these temperatures, the only other phase observed was hematite. Thus, this must have been the catalytically active phase. At lower temperatures, γ-Fe_2O_3 (maghemite) was present, however this temperature region was not sufficiently studied to conclude anything about its activity. Maghemite appeared to be the first coarsely crystalline phase formed from the nanoparticles as evidenced by its minor occurrence in the starting material. It probably grew, however, from a dissolution/re-precipitation mechanism, since it is coarser than the starting material and seems to fill void spaces. It is notable that the magnetite formed from reducing conditions when not mixed with biomass at high temperatures did not form pseudomorphs of hematite nor had entrained particles of hematite within them. It is possible that under reducing conditions the hydroxylated catalyst either converts directly to magnetite or possibly converts through an intermediate spinel (*i.e.* maghemite) phase.

24.4. CONCLUSIONS

NANOCAT® was shown to consist mostly of three components; a glass, an hydrated/hydroxylated iron oxide, and a very minor amount of maghemite, by HRTEM, electron diffraction, EDS, FTIR, EELS, and EXAFS. The composition of this material lies between that of goethite and bernalite. The material converted on heating to one of the anhydrous iron oxide phases depending on the conditions of the experiments. NANOCAT® was shown to be a much better CO catalyst and CO oxidant than other micron size iron oxides. As a catalyst, the reaction order was first order for CO and zero order for oxygen. The apparent activation energy was 14.5 Kcal/mol. The small particle size of NANOCAT® contributed to its high efficiency as a CO catalyst once the catalyst was activated, however, it was the FeOOH phase present in NANOCAT® that contributed to its low activation temperature. Therefore, the combined small particle size and FeOOH phase made NANOCAT® a superior CO catalyst for some one-time use applications. In the absence of oxygen, NANOCAT® was an effective CO oxidant, as it can directly oxidize CO to CO_2. In addition, during the direct oxidation process, the reduced form of NANOCAT® catalyzed the disproportionation reaction of CO, producing carbon deposits, iron carbide, and CO_2. The disproportionation reaction of CO contributed significantly to the total removal of CO.

Oxidative pyrolysis data of the biomass model compounds obtained from the MBMS and factor analysis indicated that NANOCAT® substantially affected the product distribution. The catalytic activity of NANOCAT® was dependent upon the model compound composition and reaction conditions used. The catalytic effects of NANOCAT® on biomass decomposition were more obvious at temperatures of 350 °C and above. Therefore, for most experiments, NANOCAT® was pre-heated to 350 °C before the reaction began. At that temperature, the FeOOH phase present in NANOCAT® had lost water and been converted to the α-Fe_2O_3 phase. The catalytic activity of FeOOH phase on biomass decomposition could not be observed. Studies by HRTEM and other spectroscopic techniques suggested that the freshly formed α-Fe_2O_3 phase of NANOCAT® may be responsible for its catalytic activity on biomass. Although this phase is predicted from classical thermodynamics data in oxidizing environments, it is not always produced in catalytic experiments with biomass.

ACKNOWLEDGEMENT

The authors gratefully acknowledge Dr. Bruce Ravel of the Naval Research Laboratory and Dr. Shaheen Islam of Virginia Union University for the XAFS analysis. Jan Lipscomb of Philip Morris was responsible for the FTIR data. The authors also thank Shahryar Rabiei and Felecia Logan for assistance in experimental works in flow tube reactor and MBMS, and Dr. Hoongsun Im for helpful discussions. Philip Morris USA management supported the work.

REFERENCES

1.   G.P. Huffman, B. Ganguly, J. Zhao, K.R.P.M. Rao, N. Shah, Z. Feng, F.E. Huggins, M.M. Taghiei, F. Lu, I. Wender, V.R. Pradhan, J.W. Tierney, M.M. Seehra, M.M. Ibrahim, J. Shabtai, and E.M. Eyring, Structure and dispersion of iron-based catalysts for direct coal liquefaction, *Energy Fuels* 7, 285-296 (1993).

2.   J. Zhao, F.E. Huggins, Z. Feng, F. Lu, N. Shah, G. P. Huffman, Structure of a nanophase iron oxide catalyst, *J. Catal.* 143, 499-509 (1993).

3.   F. Zhen, J. Zhao, F.E. Huggins, G.P. Huffman, Agglomeration and phase transition of a nanophase iron oxide catalyst, *J. Catal.* 143, 510-519 (1993).

4.   J.S. Walker, G. I. Straguzzi, W.H. Manogue, G.C. A Schuit, Carbon monoxide and propene oxidation by iron oxides for auto-emission control, *J. Catal.* 110, 298-309 (1988).

5.   P. Li, D. E. Miser, S. Rabiei, R. T. Yadav, and M. R. Hajaligol, The removal of carbon monoxide by iron 0oxide nanoparticles, *Appl. Catal. B* 43, 151-162 (2003).

6.   H.L. Chum, and R.P. Overend, Biomass and renewable fuels, *Fuel Process. Tech.* 71, 187- 195 (2001).

7.   W. Windig, J.L. Lippert, M.J. Robbins, K.R. Kresinske, J.P. Twist, and A.P. Snyder, Interactive self-modeling multivariate analysis, *Chemometrics and Intelligent Laboratory Systems* 9, 7-30 (1990).

8.   International Centre for Diffraction Data (ICDD) (Newtown Square, PA, 2001).

9.   A. Szytula, A. Burewicz, Z, Dimitrijevic, S. Krasnicki, H. Rzany, J. Wanic, W. Wolski, Neutro diffraction studies of α-FeOOH, *Phys. Status Solidi* 26, 429-434 (1968).

10. T. Liu, L. Guo, Y. Yao, T. D. Hu, Y. N. Xie and J. Zhang, Bondlength alteration of nanoparticles Fe_2O_3 coated with organic surfactants probed by EXAFS, *NanoStructured Materials* 11, 1329-1334 (1999).

11. Y.K. Rao, A physico-chemical model for reactions between particulate solids occuring through gaseous intermediates-I. Reduction of hematite by carbon, *Chem. Eng. Sci.* 29, 1435-1445 (1974).

12. P. K. DE Bokx, A. J. H. M. Kock, E. Boellaard, W. Klop, and J. W. Geus, The formation of filamentous carbon on iron and nickel catalysts, I. Thermodynamics, *J. Catal.* 96, 454-467 (1985).

13. P. K. DE Bokx, A. J. H. M. Kock, E. Boellaard, W. Klop, and J. W. Geus, The formation of filamentous carbon on iron and nickel catalysts, II. Mechanism, *J. Catal.* 96, 468-480 (1985).

14. J. Barkauskas and V. Samanavičiūtė, Kinetic investigation of CO disproportionation on Fe catalyst, *Catal. lett.* 71, 237-240 (2001)

15. E.J. Shin, M. Nimlos, and R.J. Evans, Kinetic analysis of the gas-pahse pyrolysis of carbohydrates, *Fuel* 80, 1697-1709 (2001).

16. R.J. Evans, and T.A. Milne, Molecular Characterization of the pyrolysis of biomass: fundamentals, *Energy and Fuels* 1(2), 123-137 (1987).

17. E.J. Shin, M. Nimlos, and R.J. Evans, A study of the mechanisms of vanillin pyrolysis by mass spectrometry and multivariate analysis, *Fuel* 80 (12) 1689-1696 (2001).

18. E.J. Shin, M. Nimlos, and R.J. Evans, The formation of aromatics from the gas-phase pyrolysis of stigmasterol: kinetics, *Fuel* 80 (12), 1681-1687 (2001).

25

Graphitic Nanofilaments: A Superior Support of Ru-Ba Catalyst for Ammonia Synthesis

C. H. Liang[*], Z. L. Li, J. S. Qiu, Z. B. Wei, Q. Xin and C. Li

25.1. INTRODUCTION

Catalytic synthesis of ammonia is one of the most important industrial processes in which a promoted iron catalyst is widely used at high temperature and high pressure.[1] The process under such reaction conditions is energy-intensive as well as capital-intensive although some advances have been made. Therefore, ammonia synthesis process with low energy consumption and low capital investment has been a longstanding target in the chemical industry. One of the keys to reaching this target is to develop catalysts with high activity and high stability under milder conditions. Ru-based catalysts for ammonia synthesis have also been studied in the early investigation carried out by Mittasch,[2] but superior activity was not found. An alkali-metal-promoted carbon-supported ruthenium catalyst was found to increase 10-fold in activity over the promoted-iron catalysts for ammonia synthesis under similar conditions by Aika and Ozaki in the early 1970s.[3] Since then, a number of promoted ruthenium-based catalysts with various supports and promoters have been studied in both fundamental understanding and industrial application by some research groups, including Aika et al.[3-6] in Japan, Tennison et al.[7] at British Petroleum, Muhler et al.[8-10] in Germany, Kowalczyk et al.[11-14] in Poland, Forni et al.[15, 16] in Italy, groups[17-20] in Denmark, and others.[21-25] A promoted ruthenium catalyst supported on graphite-containing carbon with high surface area had been introduced to produce ammonia from dinitrogen and dihydrogen in industrial scale.[26]

[*] C.H. Liang, Z.L. Li, Z.B. Wei, Q. Xin, C. Li, State Key Laboratory of Catalysis, Dalian Institute of Chemical Physics, Chinese Academy of Sciences, Dalian, China. Fax: 86-411-4694447, Email: chliang@dicp.ac.cn. J.S. Qiu, Department of Material and Chemical Engineering, Dalian University of Technology, Dalian, China.

Various supports, such as carbon materials, alumina, magnesia, zeolites and rare earth oxides were used in the promoted Ru catalysts for the ammonia synthesis. It seems that the ruthenium supported on carbon materials with promoters is the most promising catalyst for commercial application. However, methanation of activated carbon cannot be avoided during the reaction. High temperature treatment of carbon support can lead to graphitization, which was considered to be responsible for the high activity of carbon-supported Ru catalysts.[5, 7, 14, 15, 27] At the same time, high temperature treatment can also decrease methanation reaction even at the conditions above 700 °C and 100 bar,[15] which are more severe than the conventional reaction conditions on Ru-based catalysts. Recently, Chen and co-workers have tested the activity of K-Ru supported on multi-walled nanotubes for ammonia synthesis and found that the catalyst is more active than that on other supports.[28] It has been found that a Ru-Ba supported on graphitic nanofilaments (GNFs) catalyst is more active than the activated carbon (AC) supported Ru-Ba catalyst by our group.[29] In this work, a Ru-Ba/GNFs catalyst with a Ba/Ru molar ratio of ca. 0.25 was shown to exhibit a remarkably high activity and stability for ammonia synthesis. The methanation resistance of the GNFs support was investigated in ammonia synthesis. The kinetics of ammonia synthesis with the Ru-Ba/GNFs catalysts was also analyzed.

25.2. EXPERIMENTAL SECTION

GNFs were synthesized by CH_4 decomposition over a nickel-based catalyst, and then the catalyst was removed by dissolution typically in aqueous solution of HNO_3. For comparison, a commercial activated carbon (AC) was obtained from Beijing Guanghua Wood Plant (Beijing, China). BET surface area and elemental composition of GNFs and AC are listed in Table 25.1. The Ru-Ba/GNFs and Ru-Ba/AC catalysts were prepared by using two-step impregnation method.[24, 29] Namely, The Ru precursor salt, $RuCl_3$, was dissolved into acetone and the GNFs were added to form slurry. The sample was dried overnight at 110 °C, reduced at 400 °C (until no chlorine ions were detected with the solution of $AgNO_3$), and passivated in 1% O_2 for 2 hrs at room temperature. The as-prepared sample was impregnated with the solution of $Ba(NO_3)_2$, then dried overnight at 110 °C. Ru-Ba/AC catalysts were prepared with the same methods.

TABLE 25.1. BET surface area and elemental composition of GNFs and AC.

Sample	S_{BET} (m²/g)	Elemental composition (wt. %)					
		C	H	O	N	S	Cl
GNFs	140	98.76	0.02	1.22	/	/	/
AC	1290	81.79	2.07	7.75	0.04	0.15	0.10

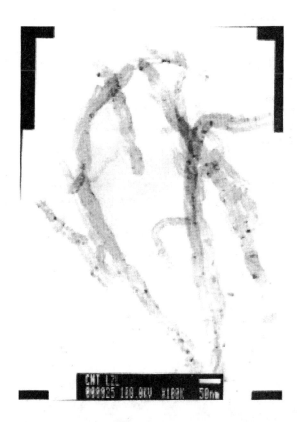

FIGURE 25. 1.   Typical TEM image of the GNFs sample.

Catalytic reactions of ammonia synthesis were carried out in a fixed-bed stainless steel micro-reactor with a stoichiometric H_2 and N_2 mixture flow at 3.0 MPa. The purity of H_2 or N_2 gases is over 99.99%, and the mixture gases were further purified before the reaction by self-designed guard containers packed with palladium catalyst and molecular sieves. Generally, the catalyst in the reactor was activated in the stream of N_2+3H_2 mixture according to the following temperature program: heating to 450 °C in 100 min, standing at 450 °C for 240 min, and then cooling to the reaction temperature in 30 min. The amount of ammonia in the effluent was determined by chemical titration method. A well-dispersed 4.0 wt. % Ru-Ba/AC catalyst was used as the reference catalyst.

The methanation of the catalysts was conducted using temperature-programmed reduction (TPR) by hydrogen coupled with a quadrupole mass spectrometer (Omnistar, Balzers Instruments). About 100 mg catalyst was first treated with He of 40 Nml/min for 1 h at 100 °C, and then the temperature was raised at a heating rate of 10 °C/min from 100 °C up to 700 °C in the presence of hydrogen.

Transmission electron microscopy (TEM) studies were carried out using a JEOL JEM-2000EX electron microscope with 100 kV.

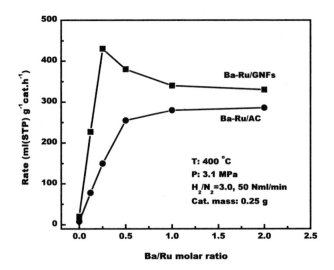

FIGURE 25.2. Comparison between ammonia synthesis activity of the Ru-Ba/AC catalysts and the Ru-Ba/GNFs catalysts with Ba/Ru molar ratio.

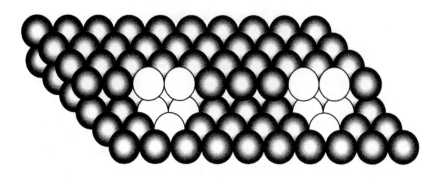

FIGURE 25.3. Schematic diagram of B_5-type sites (Two distinct sites are shown in white).

25.3. RESULTS AND DISCUSSION

Figure 25.1 shows typical TEM image of the GNFs sample. It can be seen that the GNFs sample does not contain the residues, such as Ni catalyst and its support, and few carbonaceous impurities agglomerate. The outer diameters of the GNFs were in the range of 10-40 nm, and inner diameters were in the range of 5-10 nm. The BET surface area of the GNFs, determined by nitrogen adsorption at 77 K, is about 140 m^2/g, and the BET surface area of activated carbon as reference is about 1, 290 m^2/g.

Figure 25.2 compares the ammonia synthesis activity of Ru-Ba/GNFs catalyst with Ru-Ba/AC with different Ba/Ru molar ratio. It is evident that the activity of ammonia synthesis on Ru-Ba/GNFs was improved relative to that on Ru-Ba/AC catalyst. The ammonia synthesis activity of Ru-Ba/GNFs catalyst increases when Ba/Ru atomic ratio is below 0.25, but decreases with further increasing of the amount of promoter, while that of Ru-Ba/AC catalyst increases gradually in the range of Ba/Ru atomic ratio studied. This behavior may be attributed to the low impurity in GNFs. Elemental analysis shows that AC contains some impurities such as S (0.15 %), N (0.04 %), O (7.75 %) and Cl (0.10 %), while GNFs only contains traces of oxygen (1.22 %) and hydrogen (0.02 %) besides carbon (98.76 %). The highest reaction rate of ammonia synthesis can be reached when Ba/Ru molar ratio is about 0.25 that is much lower than the optimum Ba/Ru on graphite and activated carbon supports. It is suggested that the surface impurities from carbon materials, such as S, N, O and Cl, can suppress the catalytic activity and consume part of promoter. Zhong and Aika found the removal of impurities such as S and Cl as well as acidic functional groups by hydrogen treatment led to high activity of Ru catalyst for ammonia synthesis, and believed that the elimination of the impurities species can reduce the capability of carbon support to withdraw electrons from the Ru atoms and the promoters and simultaneously enhance the electron donation of the promoter to Ru.[6]

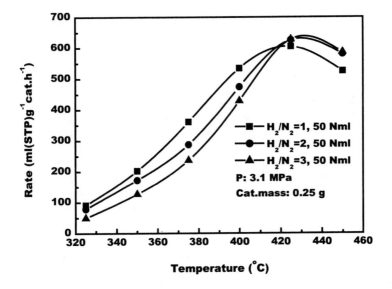

FIGURE 25.4. Ammonia synthesis rate observed for Ru-Ba/GNFs catalyst with Ba/Ru=0.25 as a function of reaction temperatures for different H$_2$/N$_2$ molar ratio at 3.0 MPa total pressure.

In addition, it has been shown that the B_5-type sites dominate the activity of the catalyst for the ammonia synthesis on the Ru catalysts. [17, 18] As shown in Figure 25.3, the B_5-type site consists of an arrangement of three ruthenium atoms in one layer and two further ruthenium atoms in the layer directly above this at a monoatomic step on an Ru(0001) terrace. [17, 18] For the unpromoted Ru catalyst, the Ru particles with about 2 nm can give the maximum number of the B_5-type sites. In the case of Ru-Ba/GNFs catalyst, the B_5-type sites may be easier to be formed than in Ru-Ba/AC catalyst as a result of the specific interaction between metal and support. However, the nature of this interaction is unclear. In the case of cinnamaldehyde hydrogenation on the Ru supported on carbon nanotubes catalyst, a specific interaction between metal and support was suggested to explain the high selectivity and high conversion. [30] Park and Baker found that the carbon nanofiber structure has an important impact on improvement of catalytic activity and selectivity on the supported Ni catalysts and believed that the observed variations in catalytic activity and selectivity can be attributed to the different orientations adopted by the nickel particles on the support materials and the graphite platelets in the nanofiber support media act as templates for the dispersed nickel crystallites, which adopt a specific geometry that is determined by their specific site location. [31] Recently, GNFs supported platinum catalysts have shown improved oxidation activity for methanol when compared to Vulcan carbon. The improvement is also believed to arise from the fact that the metal particles adopt specific crystallographic orientations when dispersed on GNFs. [32] Therefore, it is believed that the GNFs support, determining the crystallographic orientations of Ru, may play an important role in the catalytic activity of ammonia synthesis.

Figure 25.4 shows the ammonia synthesis rate observed for Ru-Ba/GNFs catalyst with Ba/Ru=0.25 as a function of reaction temperatures for different H_2/N_2 molar ratio at 3.0 MPa total pressure. It is obvious that all curves consist of a kinetically controlled region at low temperatures and a thermodynamically controlled region at high temperatures with the transition temperature lowering for lower H_2/N_2 molar ratio. It is also seen that lower H_2/N_2 molar ratio in the feed gas is favorable to get a higher activity of ammonia synthesis at low temperature, indicating that the strong adsorbed hydrogen inhibits the N_2 dissociation and activation on Ru catalysts. This result is in agreement with that obtained from other Ru catalysts. [8, 10, 19, 21] Hinrichsen et al. [9, 10] carried out kinetic simulation of ammonia synthesis catalyzed by Ru and further confirmed that hydrogen blocks active surface sites and correspondingly inhibits ammonia formation. In agreement with the results of experiments and kinetic simulation, Tennison proposed a lower H/N molar ratio for commercial operation on alternative non-iron catalysts for ammonia synthesis. [7]

The power law exponents and the apparent activation energies were derived from the analysis given by Aika and coworkers. [33] The reaction order of ammonia and the reaction orders of nitrogen and hydrogen were determined by varying the feed gas flow between 30 Nml/min and 90 Nml/min and by varying the H_2/N_2 molar ratio between 3/1 and 1/3 using a total flow of 50 Nml/min, respectively. Both determinations were carried out in the temperature range from 325 °C to 400 °C. The reaction orders of ammonia, nitrogen and hydrogen are about –0.4, 0.7 and –0.7, respectively. The power law exponents can be expressed using $r = k_{app} P_{NH3}^{-0.4} P_{N2}^{0.7} P_{H2}^{-0.7}$. The reaction orders for NH_3 and H_2 were negative, indicating that ammonia synthesis reaction is retarded the adsorbed NHx and H_2 species on Ru-Ba/GNFs catalyst. The reaction orders for N_2 was positive and close to 1

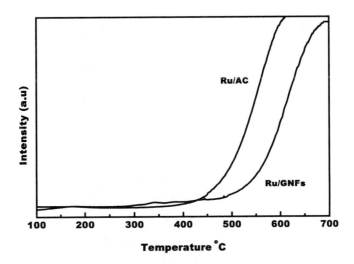

FIGURE 25.5.   TPR-MS profiles of the Ru/GNFs sample and the Ru/AC sample (methane signal, m/e = 15).

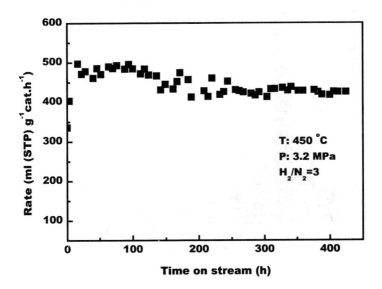

FIGURE 25.6. The stability of the Ru-Ba/GNFs catalyst at 450 °C and 3.2 MPa using a mixture gas with H_2 /N_2 =3 at a flow rate 50 Nml/min.

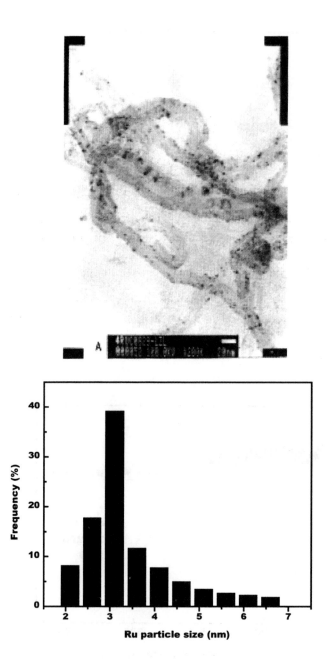

FIGURE 25.7. Representative transmission electron micrograph and Ru particle size distribution of the Ru/GNFs sample with 4.0 wt. % Ru before ammonia synthesis reaction.

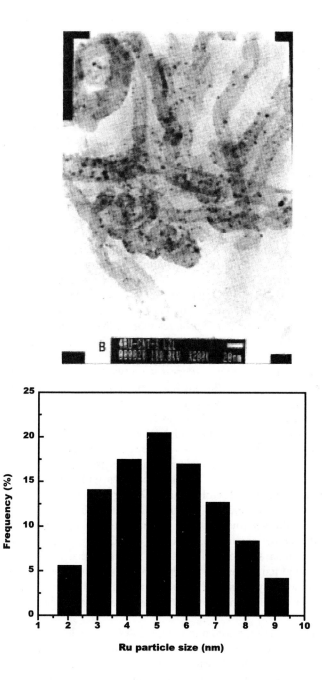

FIGURE 25.8. Representative *ex situ* transmission electron micrograph and Ru particle size distribution of a Ru-Ba/GNFs catalyst after 450 h of operation at 450 °C and 3.2 Mpa (Ru loading is about 4.0 wt % and Ba/Ru molar ratio is ca. 0.25)

indicating that the dissociative chemisorption of N_2 is the rate-determining step in ammonia synthesis. The activation energy of ammonia synthesis on the Ru-Ba/GNFs catalyst at constant flow is about 84 kJ/mol, which is close to those reported in the literatures.[11, 19, 33] Aika et al.[33] investigated kinetics of ammonia synthesis on Ru powder and oxides supported Ru catalysts with / without alkali metal promoters, and found that the reaction order in N_2 was unity for all the studied catalysts, while the reaction orders for NH_3 and H_2 were near zero and negative, respectively, for the oxides supported Ru catalysts with alkali metal promoters. They also suggested that the retarding species were both N (NH) and H over the oxides supported Ru catalysts, and H alone over Ru powder and the oxides supported Ru catalysts with alkali metal promoters.

The resistance to methanation reaction for both the Ru/GNFs sample and the Ru/AC sample was conducted using TPR-MS. The result was shown in Figure 25.5. It can be seen that methanation reaction dramatically accelerates at about 470 °C for the Ru/AC sample, while at about 550 °C for the Ru/GNFs sample. This can be attributed to the graphitic structure of GNFs. Although methanation reaction on the Ru/GNFs could not be eliminated completely, the initial temperature of the methanation reaction on the Ru/GNFs increased about 70°C compared with that of the Ru/AC, indicating that GNFs supported catalysts have higher stability for the methanation reaction than the activated carbon supported catalysts during hydrogenation reactions. Kowalczyk et al. studied the methanation of the carbon supports catalyzed by Ru catalysts for ammonia synthesis, and found that Ru readily catalyzes the methanation reaction of the carbon support.[13] Forni et al. reported that methane formation dropped dramatically at least up to 600 °C when the carbon support was treated at temperature above 1, 900 °C and suggested the graphitisation degree and purity of carbon support were two key factors for methanation reaction.[15] In the case of Ru/GNFs, the methanation resistance of Ru/GNFs can be ascribed to the graphitic structure and high purity of GNFs.

The stability of the Ru-Ba/GNFs catalyst was tested at 450 °C and 3.2 MPa using a mixture gas with H_2 /N_2 =3 at a flow rate 50 Nml/min. Figure 25.6 shows ammonia synthesis rate versus time on stream curve. It is obvious that the activity increases at the beginning of the reaction, in which the residual chlorines from Ru presursor may be slowly removed in the reaction process. The B_5-type sites with high activity may be formed under reaction condition at early stage. During test process, the ammonia concentration in the exit reached the equilibration value. No obvious deactivation was observed under those conditions after 450 hrs, while the Ru-Ba/AC catalyst shows the decreased activity under similar conditions, indicating that GNFs is a superior support for Ru-Ba catalysts of ammonia synthesis. Jacobsen[19] reported that a barium-promoted boron nitride (BN) -supported ruthenium catalyst exhibits unprecedented activity and stability in catalytic ammonia synthesis, which is due to a layered structure of BN similar to that of graphite. So the graphitic nature of GNFs may be an important factor for high activity and stability.

Representative transmission electron micrograph and Ru particle size distributions of the Ru/GNFs sample before ammonia synthesis reaction and the Ru-Ba/GNFs catalyst after ammonia synthesis reaction are shown in Figure 25.7 and Figure 25.8, respectively. In the case of Ru/GNFs, high dispersion of Ru metal particles with a narrow size distribution of 2-4 nm were seen as shown in Figure 25.7, and Ru particles mainly deposit in the outer surface of GNFs. The result is similar to that of Pt/GNFs.[32, 34] In the case of Pt/GNFs, the Pt crystals dispersed on the GNFs were relatively thin, highly crystalline faceted structure, while Pt particles supported on the activated carbon adopt a dense globular morphology.[32] Planeix and co-workers prepared Ru supported carbon nanotubes and found that Ru metal particles deposit homogeneously on the external surface of the tubes and that a high selectivity to cinnamyl alcohol was maintained in the hydrogenation of cinnamaldehyde, which is much higher than that observed with conventional Ru/C catalysts.[33] The authors suggested that a metal-support interaction could explain the very interesting observations. It is believed that the interaction between metal and support results in high activity for ammonia synthesis on the Ru-Ba/GNFs catalyst.

Figure 25.8 is a representative *ex situ* transmission electron micrograph and Ru particle size distribution of a Ru-Ba/GNFs catalyst after 450 hrs of operation at 450 °C and 3.2 MPa. It can be seen that Ru crystals and Ba compounds are about 2-9 nm and localized primarily in the outer surface of GNFs. It should be noted that the Ru particles become slightly larger after reaction. Hansen et al.[35] showed that atomic-resolution in situ transmission electron microscopy can be used to obtain insight into the structure of barium-promoted Ru/BN catalysts. The conventional *ex situ* TEM shows that all of the Ru crystals are completely covered by the support material BN, while the *in situ* TEM can show that the BN is not covering the Ru crystals, the small Ru crystals coalesce (sinter) into larger crystals, the morphology of the Ru crystals was not altered by the presence of Ba, and two distinctly new structures can be seen on the exterior of the Ru crystals in the Ba-promoted catalyst. Therefore, the bigger Ru particle may be explained by the fact that the Ru crystals are completely covered by promoter or support materials and/or sintered.

25.4. CONCLUSION

The Ru-Ba/GNFs catalyst shows remarkably high activity and stability for ammonia synthesis, which is a better catalyst than the Ru-Ba/AC. The high purity, high graphitization and unique structure of GNFs are likely to be responsible for the high activity and stability for ammonia synthesis on the Ru-Ba/GNFs catalyst. The optimized amount with Ba/Ru molar ratio for the catalytic activity is about 0.25. Kinetics analysis on the ammonia synthesis over Ru-Ba/GNFs catalyst indicates that the power law exponents can be expressed using $r = k_{app}P_{NH3}^{-0.4}P_{N2}^{0.7}P_{H2}^{-0.7}$. The apparent activation energy is 84 kJ/mol under the conditions studied. The catalyst particles become larger after reaction, probably because the Ru crystals are completely covered by promoter or support materials and/or sintered.

ACKNOWLEDGEMENTS

One of the authors (CH Liang) would like to thank Dr. Bing Zhou from Headwaters Incorporated for invaluable discussions and revision in English.

REFERENCES

1.   R. Schlögl, in: *Handbook of Heterogeneous Catalysis*, edited by G. Ertl, H. Knözinger and J.

Weitkamp (Wiley-VCH, Weiheim, 1997), pp.1697-1748.

2. A. Mittasch, Early studies of multicomponent catalysts, *Adv. Catal.* **2**, 81-104 (1950).
3. A. Ozaki, K. Aika and H. Hori, A new catalyst system for ammonia synthesis, *Bull. Chem. Soc. Jpn.* **44**(11), 3216-3216 (1971).
4. K. Aika, heterogeneous catalysis of ammonia synthesis at room temperature and atmospheric pressure, *Angew. Chem. Int. Ed. Engl.* **25**(6), 558-559 (1986).
5. H.S. Zeng, K. Inazu and K. Aika, The working state of the barium promoter in ammonia synthesis over an active-carbon-supported ruthenium catalyst using barium nitrate as the promoter precursor, *J. Catal.* **211**(1), 33-41 (2002).
6. Z.H. Zhong and K. Aika, Effect of hydrogen treatment of active carbon as a support for promoted ruthenium catalysts for ammonia synthesis, *Chem. Commun.* **13**, 1223-1224 (1997).
7. S. R. Tennison, in: *Catalytic Ammonia Synthesis*, edited by J.R. Jennings (Plenum Press, New York, 1991), pp.303.
8. H. Bielawa, O. Hinrichsen, A. Birkner and M. Muhler, The ammonia-synthesis catalyst of the next generation: Barium-promoted oxide-supported Ruthenium, *Angew. Chem. Int. Ed.* **40**(6), 1061-1063 (2001).
9. O.Hinrichsen, F. Rosowski, A. Hornung, M. Muhler and G. Ertl, The kinetics of ammonia synthesis over Ru-based catalysts, *J. Catal.* **165**(1), 33-44(1997).
10. O.Hinrichsen, F. Rosowski, M. Muhler and G. Ertl, The microkinetics of ammonia synthesis catalyzed by cesium-promoted supported ruthenium, *Chem. Eng. Sci.* **51**(10), 1683-1690 (1996).
11. W. Rarog, Z. Kowalczyk, J. Sentek, D. Skladanowski and J. Zielinski, Effect of K, Cs and Ba on the kinetics of NH3 synthesis over carbon-based ruthenium catalysts, *Catal. Lett.* **68**(3-4), 163-168 (2000).
12. Z. Kowalczyk, S. Jodzis, W. Rarog, J. Zielinski, J. Pielaszek and A. Presz, Carbon-supported ruthenium catalyst for the synthesis of ammonia: the effect of the support and barium promoter on the performance, *Appl. Catal. A* **184**(1), 95-102 (1999).
13. Z. Kowalczyk, S. Jodzis, W. Rarog, J. Zielinski and J. Pielaszek, Effect of potassium and barium on the stability of a carbon-supported ruthenium catalyst for the synthesis of ammonia, *Appl. Catal. A* **173**(2), 153-160(1998).
14. Z. Kowalczyk, J. Sentek, S. Jodzis, E. Mizera, J. Goralski, T. Paryjczak and R. Diduszko, A alkali-promoted ruthenium catalyst for the synthesis of ammonia supported on thermally modified active carbon, *Catal. Lett.* **45**(1-2), 65-72 (1997).
15. L. Forni, D. Molinari, I. Rossetti and N. Perniconi, Carbon-supported promoted Ru catalyst for ammonia synthesis, *Appl. Catal. A* **185**(2), 269-275 (1999).
16. I. Rossetti, N. Perniconi and L. Forni, Promoters effect in Ru/C ammonia synthesis catalyst, *Appl. Catal. A* **208**(1-2), 271-278 (2001).
17. C.J.H. Jacobsen, S. Dahl, P.L. Hansen, E. Törnqvist, L. Jensen, H. Topsøe, D.V. Prip, P.B. Mosenshaug and I. Chorkendorff, Structure sensitivity of supported ruthenium catalysts for ammonia synthesis, *J. Mol. Catal. A* **163**(1-2), 19-26 (2000).
18. S. Dahl, A. Logadottir, R.C. Egeberg, J.H. Larsen, I. Chorkendorff, E. Törnqvist and J.K. Nørskov, Role of steps in N_2 activation on Ru (1000), *Phys. Rev. Lett.* **83**(9), 1814-1817 (1999).
19. C.J.H. Jacobsen, Boron nitride: a novel support for ruthenium-based ammonia synthesis catalysts, *J. Catal.* **200**(1), 1-3 (2000).
20. S. Dahl, J.Sehested, C.J.H. Jacobsen, E. Törnqvist and I. Chorkendorff, Surface science based microkinetics analysis of ammonia synthesis over ruthenium catalysts, *J. Catal.* **192**(2), 391-399 (2000).
21. B.C. McClaine and R.J. Davis, Isotopic transient kinetics analysis of Cs-promoted Ru/MgO during ammonia synthesis, *J. Catal.* **210**(2), 387-396 (2002).
22. M.D. Cisneros and J.H. Lunsford, Characterization and ammonia-synthesis activity of ruthenium zeolite catalysts, *J. Catal.* **141**(1), 191-205 (1993).
23. K.S.R. Rao, S.K. Masthan, P.S.S. Prasad and P.K.Rao, Effect of barium addition on the ammonia-synthesis activity of a cesium promoted ruthenium catalyst supported on carbon-covered alumina, *Appl. Catal.* **73**(1), L1-L5 (1991).
24. C.H. Liang, Z.B. Wei, Q. Xin and C. Li, Ammonia synthesis over Ru/C catalysts with different carbon supports promoted by barium and potassium compounds, *Appl. Catal. A*, **208**(1-2), 193-201 (2001).
25. C.H. Liang, Z.B.Wei, M.F. Luo, P.L. Ying, Q. Xin and C. Li, Hydrogen spillover effect in the reduction of barium nitrate of Ru-Ba(NO_3)$_2$/AC catalysts for ammonia synthesis, *Stud. Surf. Sci. Catal.* **138**, 283-290 (2001).
26. A.K. Rhodes, *Oil Gas J.* **94** (11), 37-41 (1996).

27. X.L. Zheng, S.J. Zhang, J.X. Xu and K.M. Wei, Effect of thermal and oxidative treatments of activated carbon on its surface structure and suitability as a support for barium-promoted ruthenium in ammonia synthesis catalysts, *Carbon* **40** (14), 2597-2603 (2002).

28. H.B. Chen, J.D. Lin, J. Cai, X.Y. Wang, J. Yi, J. Wang, G. Wei, Y.Z. Lin and D.W. Liao, Novel multi-walled nanotubes-supported and alkali-promoted Ru catalysts for ammonia synthesis under atmospheric pressure, *Appl. Surf. Sci.* **180**(3-4), 328-335 (2001).

29. C.H. Liang, Z.L. Li, J.S. Qiu and C. Li, Graphitic nanofilaments as novel support of Ru-Ba catalysts for ammonia synthesis, *J. Catal.* **211**(1), 278-282 (2002).

30. J.M. Planeix, N. Coustel, B. Coq, V. Brotons, P.S. Kumbhar, R. Dutartre, P. Geneste, P. Bernier and P.M. Aiayan, Application of carbon nanotubes as supports in heterogeneous catalysis, *J. Am. Chem. Soc.* **116** (17), 7935-7936 (1994).

31. C. Park and R.T.K. Baker, Catalytic behavior of graphite nanofiber supported nickel particles. 2. the influence of the nanofiber structure, *J. Phys. Chem. B* **102**(26), 5168-5177 (1998)

32. C.A. Bessel, K. Laubernds, N.M. Rogriguez and R.T.K. Baker, Graphite nanofibers as an electrode for fuel cell application, *J. Phys. Chem. B* **105**(6), 1115-1118 (2001).

33. K. Aika, M. Kumasaka, T. Oma, O. Kato, H. Matsuda, N. Watanable, K. Yamazaki, A. Ozaki and T. Onishi, Support and promoter effect of Ruthenium catalyst III kinetics ammonia synthesis over various Ru catalysts, *Appl. Catal.* **28**, 57 (1986).

34. W.Z. Li, C.H. Liang, J.S. Qiu, W.J. Zhou, H.M. Han, Z.B. Wei, G.Q. Sun and Q. Xin, Carbon nanotubes as support for cathode catalyst of a direct methanol fuel cell, *Carbon* **40**(5), 791-793 (2002).

35. T.W. Hansen, J.B. Wagner, P.L. Hansen, S. Dahl, H. Topsøe and C.J.H. Jacobsen, Atomic-resolution in situ transmission electron microscopy of a promoter of a heterogeneous catalyst, *Science* **294**(5546), 1508-1510 (2002).

Index